AF179757

Materials at the Nanoscale

Edited by Awadesh Kumar Mallik

Published in London, United Kingdom

IntechOpen

Supporting open minds since 2005

Materials at the Nanoscale
http://dx.doi.org/10.5772/intechopen.91597
Edited by Awadesh Kumar Mallik

Contributors
Fernanda de Avila Abreu, Basma H. Al-Tamimi, Saad B. H. Farid, Norihiro Shimoi, Adrian Suarez, Jose Torres, Pedro A. Martinez, Jorge Victoria, Julio Martos, Andrea Amaro, Nurit Taitel-Goldman, Vladimir Ezersky, Anielle C.A. Almeida Silva, Amanda I. S. Barbosa, Alessandra S. Silva, Elisson A. Andrade Batista, Thaís K. de Lima Rezende, Éder V. Guimarães, Ricardo S. Silva, Noelio O. Dantas, Eliete A. Alvin, Francisco R.A. dos Santos, Samanta L.M. de Matos, Jerusa M. Maria de Oliveira, Mirella S. Vieira, Eurípedes A. da Silva Filho, Lucas Anhezini, Nilvanira D. Tebaldi, Phuoc Huu Le, Le Thi Cam Tuyen, Sheng-Rui Jian, Gulzar Ahmed Ahmed Rather, Saqib Hassan, Surajit Pal, Mohd Hashim Khan, Heshu Sulaiman Rahman, Johra Khan, Tarcisio Nascimento Correa, Igor Nunes Taveira, Rogerio Presciliano de Souza Filho

© The Editor(s) and the Author(s) 2021
The rights of the editor(s) and the author(s) have been asserted in accordance with the Copyright, Designs and Patents Act 1988. All rights to the book as a whole are reserved by INTECHOPEN LIMITED. The book as a whole (compilation) cannot be reproduced, distributed or used for commercial or non-commercial purposes without INTECHOPEN LIMITED's written permission. Enquiries concerning the use of the book should be directed to INTECHOPEN LIMITED rights and permissions department (permissions@intechopen.com).
Violations are liable to prosecution under the governing Copyright Law.

Individual chapters of this publication are distributed under the terms of the Creative Commons Attribution 3.0 Unported License which permits commercial use, distribution and reproduction of the individual chapters, provided the original author(s) and source publication are appropriately acknowledged. If so indicated, certain images may not be included under the Creative Commons license. In such cases users will need to obtain permission from the license holder to reproduce the material. More details and guidelines concerning content reuse and adaptation can be found at http://www.intechopen.com/copyright-policy.html.

Notice
Statements and opinions expressed in the chapters are these of the individual contributors and not necessarily those of the editors or publisher. No responsibility is accepted for the accuracy of information contained in the published chapters. The publisher assumes no responsibility for any damage or injury to persons or property arising out of the use of any materials, instructions, methods or ideas contained in the book.

First published in London, United Kingdom, 2021 by IntechOpen
IntechOpen is the global imprint of INTECHOPEN LIMITED, registered in England and Wales, registration number: 11086078, 5 Princes Gate Court, London, SW7 2QJ, United Kingdom
Printed in Croatia

British Library Cataloguing-in-Publication Data
A catalogue record for this book is available from the British Library

Additional hard and PDF copies can be obtained from orders@intechopen.com

Materials at the Nanoscale
Edited by Awadesh Kumar Mallik
p. cm.
Print ISBN 978-1-83968-822-5
Online ISBN 978-1-83968-823-2
eBook (PDF) ISBN 978-1-83968-824-9

For EU product safety concerns:
IN TECH d.o.o., Prolaz Marije Krucifikse Kozulić 3, 51000 Rijeka, Croatia, info@intechopen.com or visit our website at intechopen.com.

We are IntechOpen, the world's leading publisher of Open Access books

Built by scientists, for scientists

5,500+
Open access books available

134,000+
International authors and editors

165M+
Downloads

156
Countries delivered to

Our authors are among the
Top 1%
most cited scientists

12.2%
Contributors from top 500 universities

CLARIVATE ANALYTICS BOOK CITATION INDEX INDEXED

WEB OF SCIENCE™

Selection of our books indexed in the Book Citation Index
in Web of Science™ Core Collection (BKCI)

Interested in publishing with us?
Contact book.department@intechopen.com

Numbers displayed above are based on latest data collected.
For more information visit www.intechopen.com

Meet the editor

Dr. Awadesh Kumar Mallik is a ceramic engineer from the University of Calcutta, India, who is currently a materials researcher. He started growing CVD diamond at the Indian Institute of Science Bangalore laboratory during his master's thesis project for the development of vacuum tribology coatings for the Indian Space Research Organisation. He continued his scientific journey of making ceramic components for high-power electron tubes at the Council of Scientific and Industrial Research laboratory CGCRI as part of a national mission for the international fusion energy ITER program. He has executed and delivered several national and international projects. His research interests include diamond materials, thin films and coatings, CVD, PVD, tribology, bioceramics, ceramic membranes, fuel cells and batteries, ceramic powder processing, and materials science in general. He obtained a Ph.D. graduate from Jadavpur University, India, and completed an FWO postdoctoral research fellowship at Hasselt University, Belgium. He has published forty peer-reviewed articles and presented many papers at international conferences.

Contents

Preface

One nanometer is a billionth of a meter. When the size of at least one of the dimensions of orderly arranged atoms in three-dimensional space is less than 100 nm it is called a nanocrystal. Nanocrystals do not follow the law of classical mechanics as they enter the quantum realm. The reduction in size causes the surface area to increase in comparison to their volume, which helps to enhance their mechanical, thermal and catalytic properties. For example, gold nanocrystals produce different colors in their colloidal suspensions as their sizes vary, thus they can be used as color-changing cosmetics in the beauty product industry. Gold nanocrystals are already in use as labels in electron microscopy for biological samples or as drug carriers, tumor detectors, inter-cellular delivery vehicles in gene therapy, radiotherapy dose enhancers, and more. The biomedical, as well as chemistry and physics fields, benefit immensely from the numerous applications of nanocrystals. Nanodimensional rotors and motors are now being fabricated for producing power inside nanoelectromechanical systems (NEMS), integrated with nanotransistors. Atomic force microscopy (AFM) tips are used for detecting stress and vibration at the atomic level. Graphene, carbon nanotubes (CNTs), and diamond are the materials that form such nanocrystals in NEMS applications. Self-assembled monolayers (SAMs) are formed by organic molecules with their heads attaching to the substrate while functional groups like thiols and silanes are kept away from the surface. In this way, they can form two- or three-dimensional superlattices. Nanocrystals are also very important for understanding the Covid-19 pandemic. The size of the virus varies from 50 nm to 200 nm. The coronavirus is a protein membrane encapsulated nucleocapsid (N) protein and RNA genome with about 20-nm long spike proteins all around them. To fight the virus humans have to develop medicines and vaccines, which again, are made up of nanocrystals.

This book contains chapters written by researchers in biology, chemistry, physics, and engineering who study nanocrystals.

Dr. Awadesh Kumar Mallik
IMO-IMOMEC,
Hasselt University,
Diepenbeek, Belgium

Section 1

Synthesis

Chapter 1

Nonthermal Crystalline Forming of Ceramic Nanoparticles by Non-Equilibrium Excitation Reaction Field of Electron

Norihiro Shimoi

Abstract

In this work, we have discovered a method of forming ZnO thin films with high mobility, high carrier density and low resistivity on plastic (PET) films using non-equilibrium reaction fields, even when the films are deposited without heating, and we have also found a thin film formation technique using a wet process that is different from conventional deposition techniques. The field emission electron-beam irradiation treatment energetically activates the surface of ZnO particles and decomposes each ZnO particles. The energy transfer between zinc ions and ZnO surface and the oxygen present in the atmosphere around the ZnO particles induce the oxidation of zinc. In addition, the ZnO thin films obtained in this study successfully possess high functional thin films with high electrical properties, including high hole mobility of 208.6 cm^2/Vs, despite being on PET film substrates. These results contribute to the discovery of a mechanism to create highly functional oxide thin films using a simple two-dimensional process without any heat treatment on the substrate or during film deposition. In addition, we have elucidated the interfacial phenomena and crosslinking mechanisms that occur during the bonding of metal oxide particles, and understood the interfacial physical properties and their effects on the electronic structure. and surface/interface control, and control of higher-order functional properties in metal/ceramics/semiconductor composites, and contribute to the provision of next-generation nanodevice components in a broad sense.

Keywords: non-equilibrium reaction excitation field, field emission, zinc oxide, nanoparticle, conductive ceramic film

1. Introduction

As the Internet of Things (IoT) continues to grow, networks are being built in which various devices share information with each other. At present, data from various devices is scattered and siloed, but it is predicted that in the latter half of the 2020s, hundreds of billions of devices will be connected to the Internet. In the second half of the 2020s, hundreds of billions of devices are expected to be connected to the Internet. A vast amount of data and information will be constantly being formed, but if the data is siloed, it will be impossible to share it.

Therefore, it is necessary to create a wearable device that can constantly hold and share data. In order to achieve this, electronic devices with ubiquitous functions are expected to be realized as wearable devices, and flexible shapes, light weight, large area, and optically transparent functions are expected to be necessary.

Conductive oxide thin films such as indium-tin-oxide (ITO) and zinc oxide (ZnO) are expected to be the key semiconductor elements for wearable devices, but the manufacturing method to obtain conductive thin films with high electrical properties has not been established. ZnO thin films, which are introduced in this chapter, show great promise as photomechanical materials for photography [1, 2] and electrochemical and biosensor materials for various devices [3–7], and the applications of these thin films are diverse, including transparent electrodes, solar cells, and memory storage devices. When metal oxide thin films are used as optical films for liquid crystal displays and touch panels, both high transparency and low resistivity are required. At present, the aforementioned indium-tin oxide (ITO) and tin oxide are used in combination with other transparent materials. However, due to the low reserves of indium, it is expected to be depleted as a resource, and there is a need to search for alternative conductive materials, and research and development of alternative material technology has been conducted for many years. Research on ZnO has a history as long as that of ITO, and has produced many interesting results in recent years. ZnO has been grown by a number of methods, including sputtering [8–12], chemical vapor deposition [13], pulsed laser deposition [14], and wet coating [15]. each of these methods for growing ZnO introduces high conductance and high transparency properties into the film but high-performance ZnO films have yet to be fabricated using low-cost, simple processes. The low-cost process is characterized by a fast growth rate at low temperatures. However, ZnO fabrication techniques are sensitive to growth temperature, atmospheric pressure, and oxygen concentration, making it difficult to synthesize uniform ZnO films. In addition, the high dependence on the performance of the synthesis equipment has made it impossible to achieve a uniform crystal structure at low temperatures as well as a high growth rate. Therefore, we aimed to establish a simple, low-cost and stable fabrication process based on the wet coating method for growing ZnO films.

In this chapter, the author explains the need for a technological breakthrough to form oxide conductive thin films with stable conductivity on plastic films in order to utilize ZnO thin films as highly functional films for processing large amounts of data while maintaining their flexibility. The conductivity mechanism of oxide conductive materials has been inferred from indirect data such as the dependence of conductivity on oxygen partial pressure [1, 2]. In particular, ZnO, along with ITO, was first discovered as a semiconductor in the 1930s. Since then, its optical transparency in the visible range has been exploited, and research and development on its optical and electrical properties have been conducted for use as a transparent conductive film in memory devices, photovoltaics, transistors, and other applications. Furthermore, ZnO has a high carrier electron concentration, which is due to the potential presence of crystal defects caused by dopants and contaminants intervening in the crystal during ZnO synthesis. The problem of the electron mobility limit of ZnO has been clarified to some extent by a great deal of research and development into methods of ZnO synthesis and optimal selection of dopants. Although some progress has been made with ZnO thin films, they still do not have the high-performance characteristics to withstand the mega-data processing that will be required in IoT devices. The decrease in mobility is attributed to grain boundary scattering and ionized impurity scattering depending on the grain size, and the control mechanism of electrical properties differs depending on the synthesis form, such as thin film or nanowire.

Therefore, the author focused on the "non-equilibrium reaction field" as a bottom-up architecture. Although many studies have been reported on the control of one-dimensional nanowire growth and the physical properties of the composites by moving the atomic and molecular groups constituting the oxide particles without heating, no technology has been found to create thin-film nanocomposite structures in a simple two-dimensional process. In this study, the author succeeded in discovering the basis of this technology by using an electron source that emits electrons uniformly in a plane. In addition, we will elucidate the interfacial phenomena and crosslinking mechanisms that occur during the bonding of metal oxide particles, and clarify the effects on the physical properties of the interface, mechanical properties, optical properties, and electronic structure. In addition, the author aims to create low-dimensional nanomaterials based on heterogeneous oxide-metal bonding with various dimensions, construct hybrid structures, control nanostructures, surfaces and interfaces, and control higher-order functional properties in metal-ceramics-semiconductor composites, and provide next-generation nanodevice components in a broad sense.

2. Experiment

2.1 Nanoscale ZnO particle synthesis

Zinc nitrate ($Zn(NO_3)_2$) (high purity chemistry), ammonium carbonate ($(NH_4)_2CO_3$) (high purity chemistry), ethanol, and deionized water are used. Deionized water is purified to high purity water using a distillation vessel, and 1.0 M zinc nitrate and ammonium carbonate are dissolved in high purity water respectively. The zinc nitrate solution is dropped into the strongly stirred ammonium carbonate solution and reacts in a molar ratio of 1:1.5 (= $Zn(NO_3)_2$: $(NH_4)_2CO_3$) to synthesize a white precipitate. The precipitate was filtered and cleaned several times with high-purity water and ethanol, and then dried at 100 °C in air for 6 hours to form the precursor of zinc oxide. The ZnO particles were then sintered at 400 °C for 4 hours in an electric furnace to obtain ZnO particles [16–18]. The obtained ZnO particles had a median diameter of 0.18 μmϕ, with 89% of the synthesized particles distributed between 0.12 and 0.25 μmϕ. The powder is represented in **Figure 1**.

Figure 1.
ZnO nanoparticles.

2.2 Thin film formation

Prepare a coating of ZnO particles with zinc complex (zinc acetylacetonate (Zn(AcAc)), propanol (2-propanol) and a small amount of surfactant (thiol). See **Table 1** for the weight ratio; the ZnO particles were agglomerated and thiol was used to disperse them. The ZnO particles were agglomerated and thiol was used to disperse them. For the mixed solution of ZnO particles and Zn(AcAc) prepared in this study, the location of Zn(AcAc) in the residual solution after filtration of ZnO particles was not confirmed by Fourier Transform Infrared Spectroscopy (FT-IR). Zn(AcAc) was adsorbed on the zinc oxide surface. It is expected that Zn(AcAc) is electrostatically adsorbed on the surface of particles in the paint and the complexes serve as seeds for cross-linking between ZnO particles and promote the formation of ZnO thin films in two dimensions [19]. No solvent is added to act as a binder, and the coating is used to form a thin film. Plastic films with high transmittance, such as polyethylene terephthalate (PET) or polycarbonate film (PC), can be used as substrates to support the thin film. For thin film formation, a spray method was used, and a thin film with a thickness of 80 μm was applied before drying. The thin film was then dried using dry air at 60 °C to volatilize the solvent (propanol) and form a foil with only solids. At this point, the foil formed on the substrate is an aggregate of fine particles and almost no adhesion has occurred.

2.3 Field emission electron beam from hc-SWCNTs as a non-equilibrium excitation reaction field

In order to realize uniform planar electron emission as a non-equilibrium reaction field, it is essential to employ highly crystalline single-walled carbon nanotubes (hc-SWCNTs) in electrical devices as field emitters, and the effect of increasing the crystallinity of SWCNTs on electrical properties has been investigated [20, 21]. We employed hc-SWCNTs as field emission (FE) electron sources with a weight density of less than 1.3 mg/cm^2 and fabricated electrodes on graphite plates by electrostatic coating method. The FE electron source as the cathode electrode to induce FE was then grooved by physically carving the coating film to expose the hc-SWCNTs as the field emitter [20]. The electron source was stacked with an FE cathode fabricated with hc-SWCNTs to emit electrons, a gate electrode to block the FE electrons, an accelerator to adjust the acceleration energy of the FE beam from 0 to 120 kV, and an anode electrode placed on the ZnO particles, as shown in **Figure 2a**. Ever since Rinzler et al. predicted the possibility of electronic devices using CNTs in their landmark paper [22], carbon nanostructures with one-dimensional shapes have attracted much attention for their applications in electronic and electrical devices. The hc-SWCNTs have excellent physicochemical properties. Tohji et al. succeeded in synthesizing pure hc-SWCNTs by purification and annealing at high temperature of 1473 K and low pressure of less than 10^{-5} Pa [23]. We also established a method to evaluate the crystallinity of SWCNTs by using cryogenic thermal desorption spectroscopy and high-resolution transmission electron microscopy (HR-TEM, Hitachi High-Technologies) [24]. The HR-TEM image of such a crystal is shown in **Figure 2b**. We succeeded in uniformly dispersing hc-SWCNTs [23], which have

	ZnO	Thiol	Zinc acetylacetonate	Solvant
Composition ratio of mixture (wt%)	16.9	4.3	6.8	72.0

Table 1.
ZnO thin film formation condition in this study.

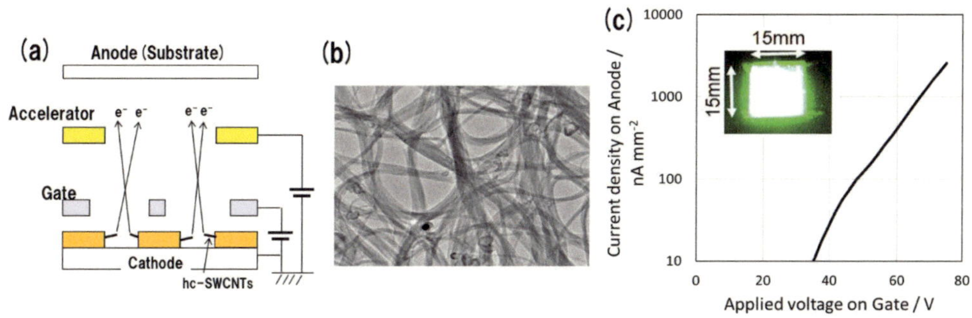

Figure 2.
FE source using hc-SWCNTs. (a) Structure of FE source with high energy electron beam to fabricate ceramic nanostructure. (b) HR-TEM image of hc-SWCNTs after crystallization. (c) FE characteristics between cathode and anode electrodes and image of lighting emitted by a planar FE cathode (inset).

excellent physicochemical properties, and they have been employed as field emitters to produce uniform planar FE electron emission over a wide range of electron emission currents, as shown in **Figure 2c**. It is difficult to form ZnO particles with uniform morphology and crystal structure when the variation of the dose in the plane is more than 0.5%. Controlling the intensity of the electron beam in the plane is essential for fabricating devices that can serve as electron sources.

It is speculated that the nanoscale Zn grains rotate and vibrate when irradiated with electrons of up to 120 keV using the system shown in **Figure 2a** [25], and that each Zn terminal interface after electron irradiation is activated as a non-equilibrium reaction field of appropriate energy of about 120 keV to bridge to other Zn atoms.

The possibility of using plastic films as substrates for general-purpose construction of key semiconductor elements for IoT devices may be well considered in the future. In order to investigate the process of device construction without heating considering the heat resistance of plastic, this study aims to form a two-dimensional continuous thin film by applying a non-equilibrium reaction field and using high-energy electron beam irradiation as the non-equilibrium reaction field.

2.4 Non-thermal formation of ZnO film on plastic film substrate by FE electron beam irradiation

We have developed a technique to fabricate ZnO thin films on thin PET substrates using FE electron beam irradiation as a non-equilibrium reaction field as shown in **Figure 3a**. The ZnO thin films were prepared by wet coating and electron beam irradiation methods [26, 27]. In this process, the thickness of the ZnO film was controlled from 100 nm to 10 μm. The FE current density emitted from each hc-SWCNT as an electron emission source was controlled to about 20 pA/cm^2, which corresponds to 10^{13} electrons/cm^2 -s for the preparation of the films. The electrical and optical properties of carrier density, hole mobility, resistivity and optical transmittance at 550 nm wavelength were evaluated for ZnO thin films of about 100 nm thickness fabricated by using 120 keV FE electron beam as basic properties. 120 keV FE electron beam irradiated ZnO thin films showed carrier density The ZnO thin films prepared by irradiation of 120 keV FE electron beam showed carrier density of 1.8 × 10^{18} cm^{-3}, hole mobility of 158.6 cm^2/Vs, resistivity of 8.6 × 10^{-4} Ω-cm and optical transmittance of 78% at 550 nm wavelength. Although attempts to improve the electrical properties of ZnO thin films using various synthetic techniques have been reported, the fabrication of highly functional

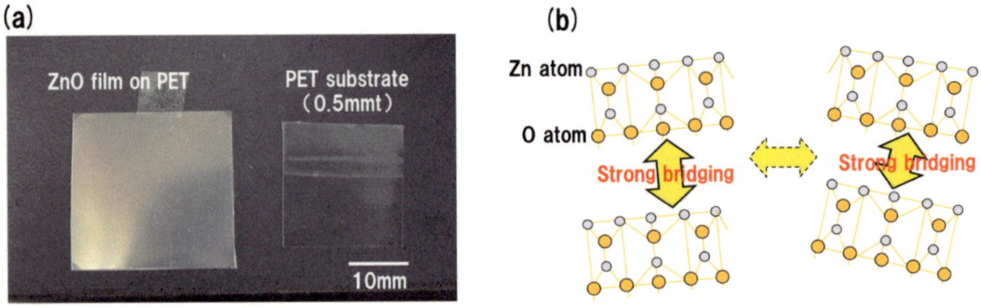

Figure 3.
(a) Images of (left) ZnO thin film synthesized on a PET substrate with irradiation of electron energy of 120 keV and (right) a bare PET substrate. (b) Schematic diagram showing bridging of ZnO crystal.

ZnO thin films on plastic films at low temperatures has not yet been done due to the temperature limitations of the substrate supporting the films. Considering the defects that stabilize the sites between the anti-sites, it is important to (1) clarify the interfacial phenomena and bridging mechanisms that selectively occur in the c-axis direction during the bonding of the terminal crystal interface of metal oxide particles as shown in **Figure 3b**, and (2) determine how to fabricate stable ZnO thin films on plastic substrates due to the constraints related to the temperature of the substrate supporting the thin films, (2) electron irradiation and acceleration energy of electron beam are necessary for crystal structure rearrangement and film synthesis after ZnO grain formation. In the present work, we propose to form ZnO bulk by FE electron beam irradiation during bridging between ZnO grains and to fabricate ceramic thin films without heat treatment.

By using non-equilibrium reaction fields as a bottom-up tool for the formation of oxide thin films, we have discovered a process technology for the formation of ZnO thin films on PET film substrates. The thin films obtained by this process have been evaluated for their conductive properties. In this study, we have evaluated the basic physical properties of this thin film such as carrier density, hole mobility, resistivity and optical transmittance. The results are shown in **Figure 4** for carrier density, hole mobility and resistivity, and **Figure 5** for transmittance [26]. We have succeeded in finding that each property differs depending on the electron beam irradiation energy. ZnO thin films grown at electron energy of 120 keV have electrical properties of carrier density of 1.8×10^{18} cm^{-3} and hole mobility of 208.6 cm^2/Vs, and resistivity of 8.6×10^{-4} Ω-cm. However, due to the limitations of the ZnO (plus dopant) formation temperature and the substrate on which the thin film is supported, low-temperature ZnO formation has not yet yielded highly functional thin films in terms of hole mobility and resistivity. O atoms, and

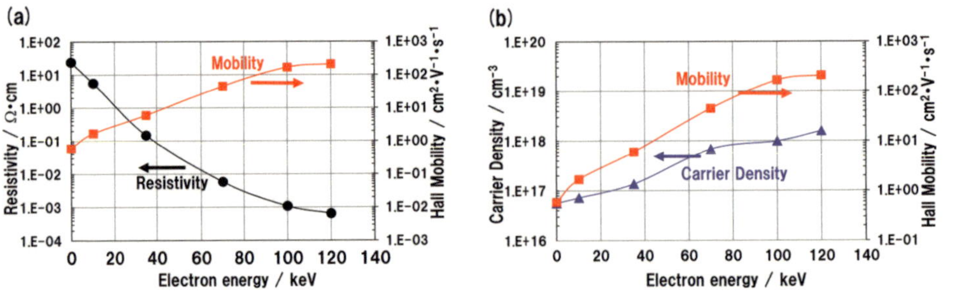

Figure 4.
(a) Resistivity and hole mobility dependence on electron beam irradiation energy, (b) carrier density and hole mobility dependence on electron beam irradiation energy.

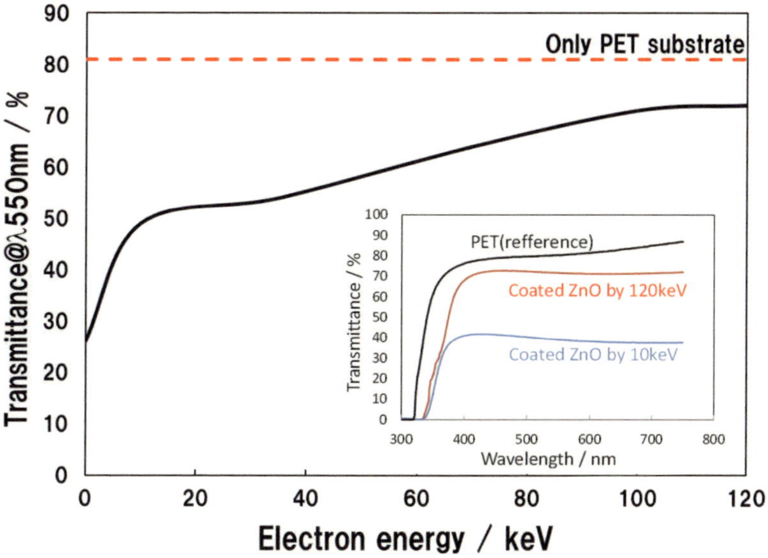

Figure 5.
Spectral transmittance of ZnO thin film on PET substrate as a function of electron irradiation energy at 550 nm. The red dotted line in the graph shows the transmittance at 550 nm for PET substrate only, and the inset shows the spectral transmittance for PET substrate only at electron irradiation energy of 10 keV (blue solid line) and 120 keV (red solid line).

complex defects that stabilize Zn interstitials and Zn anti-sites, the technology to build stable ZnO thin films or bulk structures has not been established. It has been reported that strict and stable control of the deposition environment is necessary from the time of ZnO formation with regard to oxygen defects and defects in the crystal concept, and that rearrangement of the crystal structure and control of crystal defects by annealing at 500 K or higher after the formation of the thin film are necessary to improve electrical properties [28, 29]. In our proposal, a thin film is formed by Zn cross-linking between ZnO particles and Zn partial oxidation by controlling the oxygen atmosphere during film formation without using a heating process, and we assume that the cross-linked Zn plays the role of atoms close to n-type dopants. As for the transmittance (**Figure 5**), it shows the dependence of transmittance on electron beam irradiation energy at 550 nm. The ZnO thin film has a band gap of about 3.7 eV and was found to have a crystalline structure similar to the bulk structure [29, 30].

The transmittance of the ZnO thin film was 70% @ λ550nm, the same as that of the PET substrate, and **Figure 5** shows that there is still room for improvement in the transmittance.

3. Discussion

3.1 Crystallization process of ZnO particles

The surface morphology and composition distribution of ZnO particles on PET film substrates were observed by Scanning Electron Microscope (SEM; Hitachi High-Technologies Co., Ltd., Japan) and Electron Probe Micro Analyzer (EPMA; JEOL). **Figure 6a** shows the surface morphology of the coating during the formation of ZnO thin film, and **Figure 6b** and **c** show the surface morphology of the coating after plasma treatment. **Figure 6b–e** show the compositional separation of ZnO particles as oxide and Zn (including Zn ions) as metal, analyzed by EPMA.

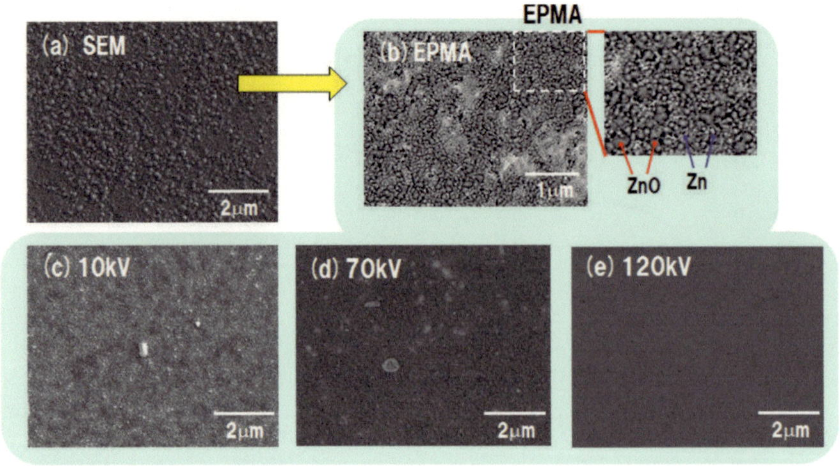

Figure 6.
The surface after ZnO coating (a) SEM after plasma treatment, (b) EPMA of the film surface after plasma treatment, (c) EPMA of the film surface after 10 keV electron beam irradiation, (d) EPMA of the film surface after 70 keV electron beam irradiation, (e) EPMA of the film surface after 120 keV electron beam irradiation.

Figure 6b shows the uniform dispersion of ZnO particles and Zn ions in submicron order and the close relationship between ZnO particles (gray) and Zn ions (white) in the coating film after plasma treatment. Ions (white) can be confirmed. Furthermore, **Figure 6c–e** show the EPMA images of the surface after irradiation with 10, 70 and 120 KeV electron beams. As the electron beam energy intensity increased, the distribution of Zn ions, represented by white color, disappeared and was found to be dominated by ZnO. However, it is difficult to form ZnO thin films with uniform composition and crystal structure when the in-plane dose density variation is more than 0.5%, and the control of the in-plane electron dose is a very important performance in the construction of electron source devices.

The electron diffraction patterns of (a) and (e) are shown in **Figure 7** using a Transmission Electron Microscope (TEM; JEOL). For the thin film immediately after coating, it represents a mass of submicron sized ZnO particles, i.e., polycrystalline structure, and it can be seen that there are individual particles with aligned crystallinity in random directions. In the crosslinking between ZnO particles, diffraction grating images along the C axis were observed, and the internal crystalline structure of the particles themselves was retained in the direction along the a and b axes. In the crosslinking between ZnO particles, diffraction grating images along the C-axis were observed, suggesting that a crystal structure similar to that of ZnO was constructed between the particles in the direction along the a- and b-axes while the internal crystal structure of the particles was retained.

The macrocrystalline structure of ZnO thin film formed by electron beam irradiation was analyzed by X-Ray diffraction (XRD; Rigaku) for its dependence on electron beam irradiation energy (**Figure 8** [26]). It was found that the grain size of ZnO-derived single crystals increased with increasing electron beam energy, indicating that the electron beam irradiation energy promoted grain cross-linking, i.e., crystal growth. Adsorbed on the ZnO particles were observed to cross-link with ZnO while aligning the crystal direction by plasma exposure and electron beam irradiation.

We were able to obtain different electrical properties depending on the process conditions of the non-equilibrium reaction field without changing the materials that make up the ZnO film. The change in the electrical properties of ZnO, especially the mobility (carrier density), is mostly due to the scattering of electrons and holes in the ZnO crystal structure by ionized impurities or grain boundaries.

The degree of scattering is greatly influenced by the control of crystallinity. In this study, only ZnO particles and zinc complexes were used for ZnO thin film formation, and we tried to analyze the surface state of ZnO to investigate factors other than the effect of dopants.

Figure 9 shows the results of X-ray Photoelectron Spectroscopy (XPS; Bruker) analysis of ZnO particles only. The binding energy of Zn intercalated in the

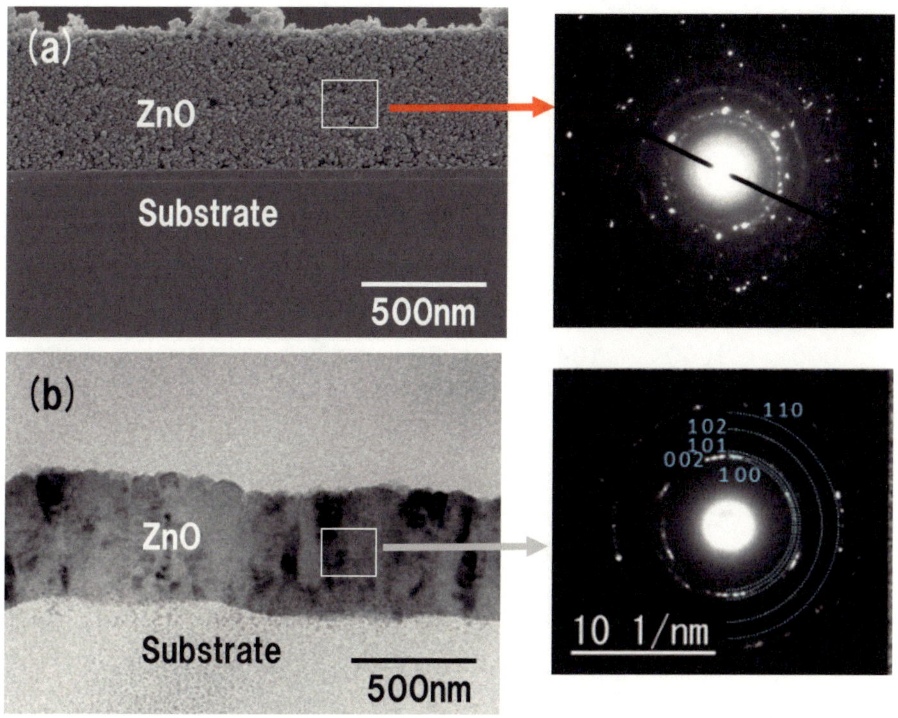

Figure 7.
(a) SEM observation and TEM diffraction image of the film cross-section after coating process, (b) TEM observation and diffraction image of the film cross-section after 120 keV electron beam irradiation.

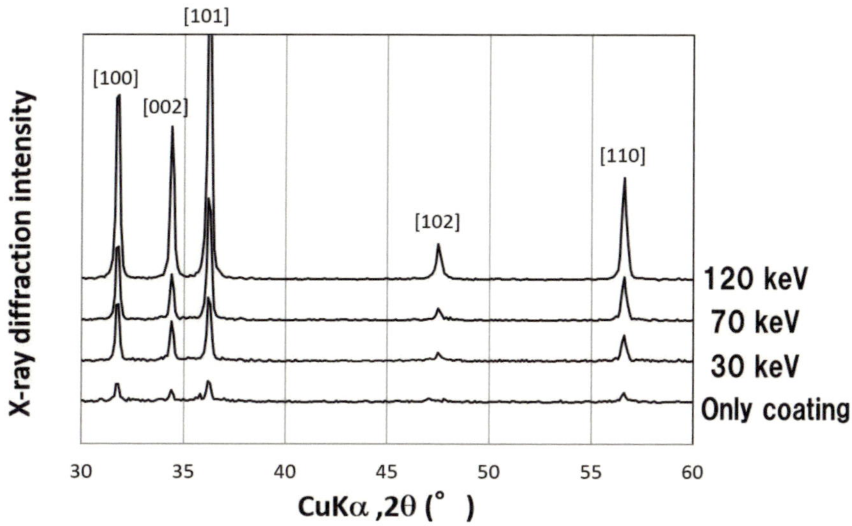

Figure 8.
XRD patterns of ZnO thin films after plasma treatment and electron irradiation.

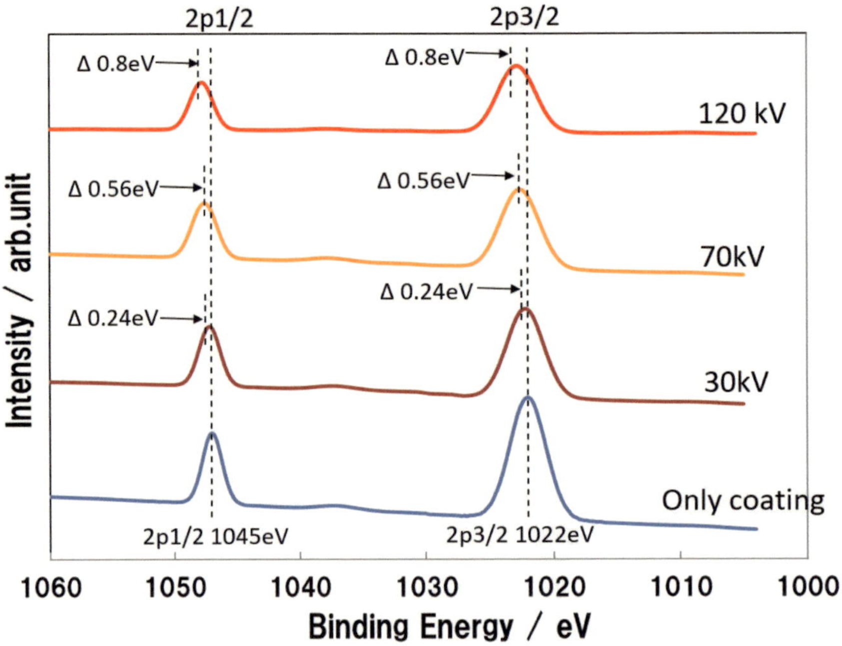

Figure 9.
Binding energy transitions of Zn 2p3/2 and 2p1/2 after plasma treatment and electron beam energy of 30, 70 and 120 keV irradiation of ZnO thin films by XPS.

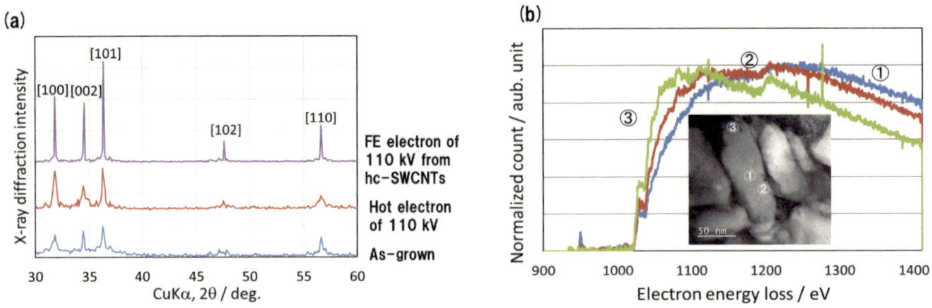

Figure 10.
Zinc spectrum of a ZnO particle synthesized by electron beam irradiation. The inset shows the measured points on the particle.

particle surface was evaluated [26]. The ZnO particles with Zn ions adsorbed from Zn(AcAc) from plasma treatment were irradiated with electron beam and it was found that the binding energy of Zn2p3/2 and Zn2p1/2 shifted to higher energy side with the increase of electron beam energy. In particular, when the electron beam energy is 110 keV, the binding energy is increased by 0.8 eV. Due to the increase in binding energy, the valence electron density of states decreases, i.e., the oxidation number increases. It is speculated that electron beam irradiation makes Zn and ZnO particle surfaces more susceptible to oxidation, and that defects such as oxygen defects and interstitial zinc on ZnO surfaces are oxidized and bound.

Figure 10 shows the spectra of zinc in ZnO particles synthesized by 120 keV electron beam irradiation, obtained from electron energy loss spectroscopy (EELS; Gatan Corporation) [31]. The measurement points were the edges of the short and long axes and the center of the ellipsoidal particles, as shown in **Figure 10b**. The spectra obtained at both edges of the particle (point 2 and point 3 in the inset of **Figure 10b**)

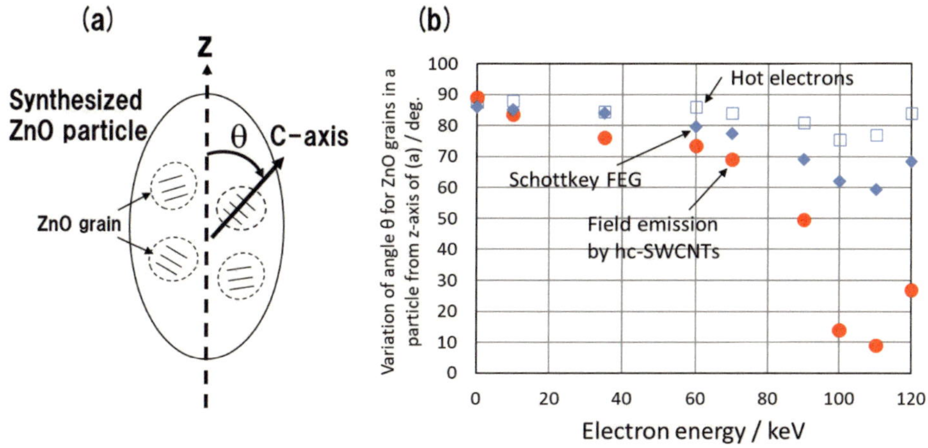

Figure 11.
Dependence of the c-axis direction and the angle from the major axis of the ZnO particles observed after electron irradiation at acceleration voltages from 0 to 120 keV. (a) Schematic diagram showing the angle between the long axis of the synthesized particles and the c-axis direction of each ZnO particle in the particles. (b) Schematic diagram showing the dependence of the angle change from the c-axis direction on the electron energy of the electron beam.

were different from those obtained at point 1 in the inset of **Figure 10b**. The spectrum obtained at point 1 provided information about the ZnO bulk. The zinc terminal interfaces of the ZnO particles synthesized by electron irradiation had different compositions and complex electrical states, and the interface belonging to the long axis of the particles (c-axis of ZnO) was found to activate the crystal bridging [32–35].

3.2 Analysis of crystal orientation of synthesized ZnO particles

The angle between the long axis, meaning the direction of the elongated ZnO particles aggregated with nanoscale ZnO particles in the synthesized particles, and the c-axis of each ZnO particle in the synthesized particles is shown in **Figure 11a**, and the angular distribution along the long axis of the particles as shown in **Figure 11b** was obtained by TEM [31]. **Figure 11b** shows the variation dependence of the angle obtained by clockwise rotation between the c-axis of each ZnO particle in the ZnO composite particles and the z-axis in **Figure 11a** on the electron energy of each electron source. The range of the angle is from 0 to 90 degrees. The results show that the c-axis of each ZnO grain in the particle is aligned with high energy resolution and high acceleration energy of about 90–120 keV. In particular, the crystallization of ZnO particles by FE electron beam irradiation resulted in fine bridges between the c-axes of each formed grain. For the electron-beam treated ZnO particles, aggregation of nanoscale ZnO particles, i.e., polycrystalline structure, was observed with crystallographically continuous structure along the c-axis similar to that of bulk ZnO crystals, indicating the distribution of individual nanoscale grains with an ordered crystal structure. The energy resolution of the electron beam depends on the electron emission mechanism.

4. Conclusion

We have developed a method for non-thermal cross-linking of ZnO particles by FE electron beam irradiation as a non-equilibrium reaction field. In order to enhance the crystallinity of ceramic materials, it is necessary to control the energy resolution and acceleration of the electron beam. In this study, an FE electron beam source with a low energy resolution and a half-value width of less than 100 meV was employed. by irradiating FE electron beams with a dose of less than 0.5% in the plane, the energy of the terminal crystalline interface of ZnO nanoscale particles was activated and the zinc ions decomposed from the ZnO particles where bridging was achieved. Thus, by energy exchange between each ZnO particle, the zinc was connected to the zinc or oxide of other site atoms. Furthermore, the processing architecture obtained in this study allowed us to fabricate ZnO thin films with good electrical properties on PET film substrates. In the future, we will aim to obtain higher electrical and optical properties by synthesizing ZnO thin films on PET film by controlling the adhesion between ZnO and substrate and the crystal orientation of the ZnO layer near the substrate. If we can control the crystal orientation and nanostructure of the ceramic film near the substrate, we expect to be able to fabricate ceramic thin films (layers) on top of other materials, semiconductors, or other ceramic layers.

This bottom-up architecture will also enable us to (1) determine the interfacial properties and their effect on the electronic structure, and (2) control the nanostructured end-crystalline interface to provide the basic elements for a wide range of next-generation devices by using composites of metals, ceramics, and semiconductors to achieve highly functional properties.

Author details

Norihiro Shimoi
Department of Electrical and Electronic Engineering, Tohoku Institute of Technology, Sendai, Japan

*Address all correspondence to: n-shimoi@tohtech.ac.jp

IntechOpen

© 2021 The Author(s). Licensee IntechOpen. This chapter is distributed under the terms of the Creative Commons Attribution License (http://creativecommons.org/licenses/by/3.0), which permits unrestricted use, distribution, and reproduction in any medium, provided the original work is properly cited.

References

[1] T. Minami, "Transparent and conductive multicomponent oxide films prepared by magnetron sputtering." J. Vac. Sci. Technol. A17, 1765 (1999).

[2] J. Suskia, D. Largeaua, A. Steyer1, V. D. Polb and F.R. Blomb, "Optically activated $ZnO/SiO_2/Si$ cantilever beams." Sens. Actuat. A: Physical, 24, 3, 221-225 (1990).

[3] S. Masuda, K. Kitamura, Y. Okumura, S. Miyatake, H. Tabata and T. Kawai, "Transparent thin film transistors using ZnO as an active channel layer and their electrical properties." J. Appl. Phys. 93, 1624 (2003).

[4] Y. Furubayashi, T. Hitosugi, Y. Yamamoto, K. Inada, G. Kinoda, Y. Hirose, T. Shimada, and T. Hasegawa, "A transparent metal: Nb-doped anatase TiO_2." Appl. Phys. Lett. 86, 252101 (2005).

[5] C.-Y. Lee, Y.-T. Haung, W.-F. Su, and C.-F. Lin, "Electroluminescence from ZnO nanoparticles/organic nanocomposites." Appl. Phys. Lett. 89, 231116 (2006).

[6] H. Kawazoe, M. Yasukawa, *H.* Hyodo, M. Kurita, H. Yanagi, and *H.* Hosono, "P-type electrical conduction in transparent thin films of $CuAlO_2$." Nature, 389, 939 (1997).

[7] P.-H. Yeh, Z. Li, and Z. L. Wang, "Schottky-Gated Probe-Free ZnO Nanowire Biosensor." Adv. Mater. 21, 4975 (2009).

[8] I.A. Rauf, "A novel method for preparing thin films with selective doping in a single evaporation step." J. Mater. Sci. Lett., 12, 1902 (1993).

[9] H. Ohta, M. Orita, M. Hirano, H. Tanji, H. Kawazoe and H. Hosono, "Highly electrically conductive indium–tin–oxide thin films epitaxially grown on yttria-stabilized zirconia (100) by pulsed-laser deposition." Appl. Phys. Lett. 76, 2740 (2000).

[10] T. Minami, T. Miyata, Y. Ohtani, and Y. Mochizuki, "New Transparent Conducting Al-doped ZnO Film Preparation Techniques for Improving Resistivity Distribution in Magnetron Sputtering Deposition." Jpn. J. Appl. Phys. 45, L409 (2006).

[11] M. Kon, P. K. Song, Y. Shigesato, P. Frachl, A. Mizukami, and K. Suzuki, "Al-Doped ZnO Films Deposited by Reactive Magnetron Sputtering in Mid-Frequency Mode with Dual Cathodes." Jpn. J. Appl. Phys. 41, 814 (2002).

[12] N. Malkomes, M. Vergohl and B. Szyszka, "Properties of aluminum-doped zinc oxide films deposited by high rate mid-frequency reactive magnetron sputtering." J. Vac. Sci. Technol. A19, 414 (2001).

[13] S. Ohnishi, Y. Hirokawa, T. Shiosaki, and A. Kawabata, "Chemical Vapor Deposition of Single-Crystalline ZnO Film with Smooth Surface on Intermediately Sputtered ZnO Thin Film on Sapphire." Jpn. J. Appl. Phys. 17, 773 (1978).

[14] R. D. Vispute, V. Talyansky, Z. Trajanovic, S. Choopun, M. Downes, R. P. Sharma, T. Venkatesan, M. C. Woods, R. T. Lareau, K. A. Jones and A. A. Iliadis, "High quality crystalline ZnO buffer layers on sapphire (001) by pulsed laser deposition for III–V nitrides." Appl. Phys. Lett. 70, 2735 (1997).

[15] *S. major*, A. Banerjee, and K. L. Chopra, "Highly transparent and conducting indium-doped zinc oxide films by spray pyrolysis." Thin Solid Films, 108, 333 (1983).

[16] O. Mahian, A. Kianifar, and S. Wongwises, "Dispersion of ZnO Nanoparticles in a Mixture of Ethylene Glycol–Water, Exploration of Temperature-Dependent Density, and Sensitivity Analysis." J. Clust. Sci. 24, 1103-1114 (2013).

[17] N. Uekawa, J. Kajiwara, N. Mochizuki, K. Kakegawa, and Y. Sasaki, "Synthesis of ZnO nanoparticles by decomposition of zinc peroxide." Chem. Lett. 30, 606-607 (2001).

[18] N. Uekawa, N. Mochizuki, J. Kajiwara, F. Mori, Y. J. Wu, and K. Kakegawa, "Nonstoichiometric properties of zinc oxide nanoparticles prepared by decomposition of zinc peroxide." Phys. Chem. Chem. Phys. 5, 929-934 (2003).

[19] N. Shimoi, T. Harada, Y. Tanaka, and S.-I. Tanaka, "Controlling the electrical properties of ZnO films by forming zinc and oxide bridges by a plasma and electron-assisted process." AIP Adv. 2, 22167 (2012).

[20] N. Shimoi, L. E. Adriana, Y. Tanaka, K. Tohji, "Properties of a field emission lighting plane employing highly crystalline single-walled carbon nanotubes fabricated by simple processes," Carbon 65, 228-235 (2013).

[21] N. Shimoi, Y. Sato, K. Tohji, "Highly Crystalline Single-walled Carbon Nanotube Field Emitters: Energy-loss-free High Current Output and Long Durability with High Power," ACS Appl. Electron. Mater. 1, 163-171 (2019).

[22] A. G. Rinzler, J. H. Hafner, P. Nikolaev, L. Lou, S. G. Kim, D. Tománek, P. Nordlander, D. T. Colbert, R. E. Smalley, "Unraveling Nanotubes: Field Emission from an Atomic Wire," Science 269, 1550-1553 (1995).

[23] K. Tohji, T. Goto, H. Takahashi, Y. Shinoda, N. Shimizu, B. Jeyadevan, I. Matsuoka, Y. Saito, A. Kasuya, T. Ohsuna, K. Hiraga, Y. Nishina, "Purifying single-walled nanotubes," Nature 383, 679 (1996).

[24] S. Iwata, Y. Sato, K. Nakai, S. Ogura, T. Okano, M. Namura, A. Kasuya, K. Tohji, K. Fukutani, "Novel method to evaluate the carbon network of single-walled carbon nanotubes by hydrogen physisorption," J. Phys. Chem. C 111, 14937-14941 (2007).

[25] B.-S. Xu, S.-I. Tanaka, "Behavior and bonding mechanisms of aluminum nanoparticles by electron beam irradiation," NanoStructured Materials 12, 915-918 (1999).

[26] N. Shimoi, S.-I. Tanaka, "Nonthermal crystal bridging of ZnO nanoparticles by nonequilibrium excitation reaction of electrons and plasma without cross-linking agent on plastic substrate," J. Alloys and Comp. 797, 676-683 (2019).

[27] N. Shimoi, T. Harada, Y. Tanaka, S.-I. Tanaka, "Controlling the electrical properties of ZnO films by forming zinc and oxide bridges by a plasma and electron-assisted process," AIP Advances 2, 022167 (2012).

[28] D. C. Look, J. W. Hemsky, and J. R. Sizelove, "Residual Native Shallow Donor in ZnO." Phys. Rev. Lett. 82 (1999) 2552-2555.

[29] K Ellmer, "Resistivity of polycrystalline zinc oxide films: current status and physical limit." J. Phys. D: Appl. Phys. 34 (2001) 3097-3108.

[30] M. Xia, Z. Cheng, J. Han, M. Zheng, C.-H. Sow, J. T. L. Thong, S. Zhang, and B. Li, "Gallium ion implantation greatly reduces thermal conductivity and enhances electronic one of ZnO nanowires." AIP Adv. 4 (2014) 57128.

[31] N. Shimoi, and S.-I. Tanaka, "Nonthermal and selective crystal bridging of ZnO grains by irradiation

with electron beam as nonequilibrium reaction field," Rev. Sci. Instrum. 92, 023905 (2021).

[32] W. P. Vellinga, J. TH. M. De Hosson, "Atomic structure and orientation relations of interfaces between Ag and ZnO," Acta Mater. 45, 933-950 (1997).

[33] K. Murakami, M. Saito, E. Takuma, H. Ichinose, "ARHVTEM of Pd/ZnO heterointerface chemical structure," J. Elec. Micro. 52, 27-32 (2003).

[34] B. Meyer, D. Marx, "Density-functional study of Cu atoms, monolayers, films, and coadsorbates on polar ZnO surfaces," Phys. Rev. B 69, 235420 (2004).

[35] S. B. Sinnott, E. C. Dickey, "Ceramic/metal interface structures and their relationship to atomic- and meso-scale properties," Mat. Sci. Eng. R 43, 1-59 (2003).

Chapter 2

Biomineralization of Magnetosomes: Billion-Year Evolution Shaping Modern Nanotools

Tarcisio Nascimento Correa, Igor Nunes Taveira, Rogerio Presciliano de Souza Filho and Fernanda de Avila Abreu

Abstract

Biomineralization in the microbial realm usually gives origin to finely structured inorganic nanomaterials. Perhaps, one of the most elegant bioinorganic processes found in nature is the iron biomineralization into magnetosomes, which is performed by magnetotactic bacteria. A magnetosome gene cluster within the bacterial genome precisely regulates the mineral synthesis. The spread and evolution of this ability among bacteria are thought to be a 2,7-billion-year process mediated by horizontal gene transfers. The produced magnetite or greigite nanocrystals coated by a biological membrane have a narrow diameter dispersibility, a highly precise morphology, and a permanent magnetic dipole due to the molecular level control. Approaches inspired by this bacterial biomineralization mechanism can imitate some of the biogenic nanomagnets characteristics in the chemical synthesis of iron oxide nanoparticles. Thus, this chapter will give a concise overview of magnetosome synthesis's main steps, some hypotheses about the evolution of magnetosomes' biomineralization, and approaches used to mimic this biological phenomenon *in vitro*.

Keywords: magnetotactic bacteria, magnetosomes, magnetic nanoparticles, magnetite, magnetosome gene cluster, horizontal gene transfer, biomimetics

1. Introduction

Among everything that is known in Microbiology, magnetotactic bacteria (MTB) are known to perform one of the finest examples of a controlled biomineralization process. MTB were first observed in the late 1950s, by the medical Doctor Salvatore Bellini in the Italian city of Pavia and later described in Massachusetts by Richard Blakemore in the 1970s [1, 2]. MTB are known to align its motility axis to the geomagnetic field and use it for orientation. When observed under the light microscope, MTB present unidirectional swimming to the North or South Magnetic Poles from an applied external magnetic field (a magnet); this behavior is called magnetotaxis [3]. This behavior occurs due to the presence of magnetic

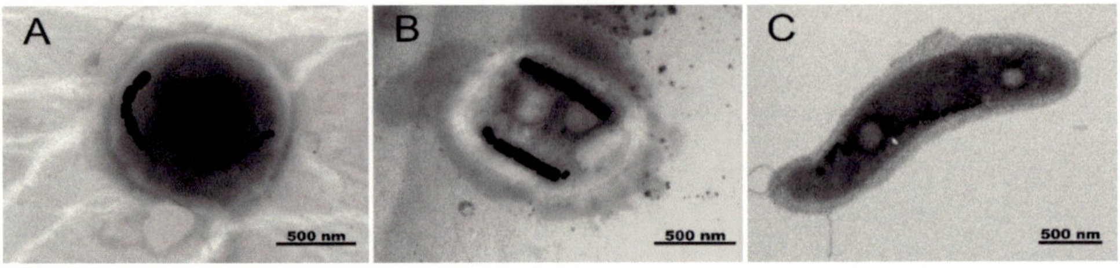

Figure 1.
Transmission electron microscopy of: (A) uncultured coccoid magnetotactic bacteria from Monsimet Cove, Antarctica. (B) Uncultured coccoid magnetotactic bacteria from Punta Ullman, Antarctica. (C) Magnetovibrio blakemorei strain MV-1^{T}.

nanocrystals—the magnetosomes—, usually aligned in single or multiple chains within the bacterial cytoplasm (**Figure 1**), and flagellar propulsion guided by chemotaxis [3]. In a simple way, chemotaxis in MTB is assisted by bacterial orientation along Earth's magnetic field (magnetotaxis). Therefore, magnetotaxis allows MTB to find the optimum position for survival and growth in a chemically stratified water column, seeking for an optimum environment where proton motive driving force reaches maximum potential. For MTB, which are frequently microaerophilic or anaerobic microrganisms, this environment is near the oxic/anoxic interface [4].

Magnetosomes are composed of a magnetic nanoparticle in most cases composed of magnetite (Fe_3O_4) and sometimes greigite (Fe_3S_4) with species specific shapes and sizes, and enveloped by a phospholipid bilayer with associated proteins, which constitutes the magnetosome membrane (MM) [3]. The gene regulation of magnetosome biomineralization (MB) and organization within the cell will be discussed in more detail in the sections ahead. Based on the total iron amount within a magnetotactic bacterium cell, MTB appear to play a major role in the biogeochemical cycling of iron [5]. MTB through magnetosome synthesis,

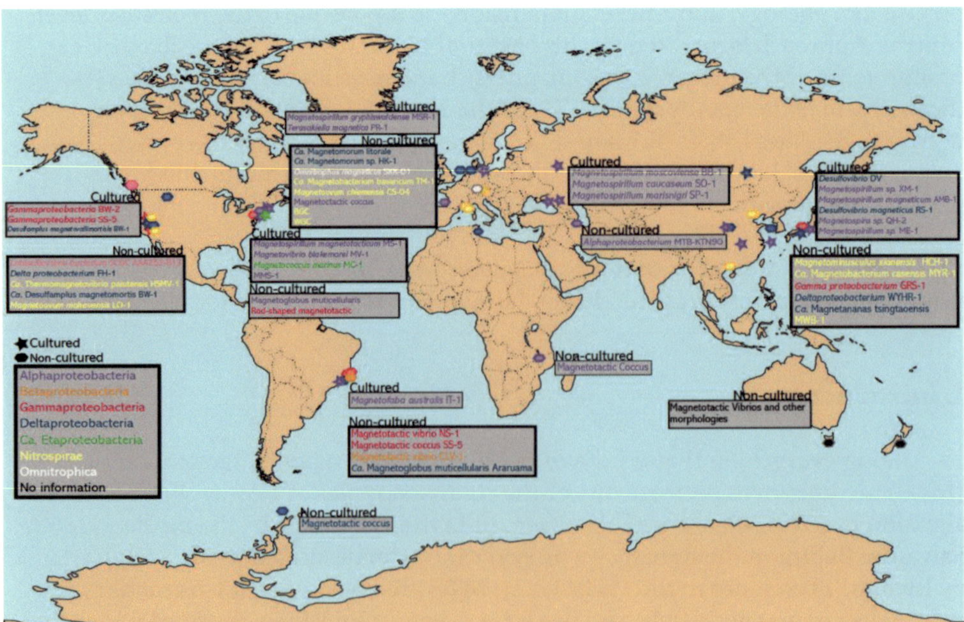

Figure 2.
Map of the distribution of known cultured and non-cultured magnetotactic bacteria across de world by phylogenetic group (see the correspondence between taxa and colors on the bottom left corner of the image).

assimilate the iron solubilized in the environment to an inorganic crystal. After cell lysis, the magnetosome is deposited in the sediment, forming what is known as magnetofossils [6]. Besides, MTB can be ingested by protozoans, and the iron from magnetosomes is, then, incorporated in the food chain [7]. Apart from iron and based on their physiology, MTB seem to have relevant roles in other biogeochemical cycles of sulfur, nitrogen, and carbon [8].

MTB are an extremely diverse group of Gram-negative bacteria with a variety of morphotypes (i.e., rods, vibrios, spirilla, coccoid, and ovoid) and species affiliated to Proteobacteria (Alpha-, Beta-, Gamma-, Delta-, and *Ca.* Etaproteobacteria class), Omnitrophica and Nitrospirae phyla [9]. MTB affiliation to other taxa have been proposed based on metagenomics studies, but observation of the magnetosomes was not performed to confirm this matter. This great diversity is reflected in MTB ubiquity in almost all aquatic habitats across the Earth (**Figure 2**), including extreme environments such as thermal trenches and saline-alkaline lakes [6, 10]. More than being interesting species for their unique evolutionary process and ecological importance, MTB are also proving to be of interest for biotechnological applications. Their unique physiology makes MTB potential bioremediators of heavy metals and magnetosomes can be extracted and used as nanotools for magnetic controlled drug targeting, contrast agents for magnetic resonance imaging, enzyme immobilization and many more industrial and biomedical applications [11].

2. Steps of magnetosome biomineralization in MTB

MB is highly regulated at the genetic level [12]. Magnetosome gene clusters (MGCs) [13], structured as *operons*, are responsible for MB in MTB. MTB genomes contain: (i) conserved *mam* genes, encountered in all MTB; and (ii) restricted genes encountered in some phylogenetic groups of MTB [14]. Examples of genes restricted to certain MTB are: (i) *mms* (from magnetosome membrane specific) genes found in magnetotactic Proteobacteria; (ii) *mad* (from magnetosome associated Deltaproteobacteria), which were first reported in magnetotactic deltaproteobacteria [15] and recently encountered in MTB affiliated to Omnitrophica and Nitrospirae phyla [9]; and (iii) *man* (from magnetosome genes in Nitrospirae), which are genes reported in MTB affiliated to Nitrospirae phylum [16]. Comprehension of MB were inferred by *mam* and *mms* genes deletion in the cultured magnetotactic alphaproteobacteria *Magnetospirillum magneticum* strain AMB-1 and *Magnetospirillum gryphiswaldense* strain MSR-1 [14]. Precise *man* and *mad* genes roles in MB remain unclear as they were studied in uncultured MTB [16], thus genetic systems to test gene function is not available.

As previously described, MTB are capable of biomineralizing magnetosomes, an organelle with a ferrimagnetic mineral core surrounded by a biological membrane [3]. A series of complex mechanisms occur in order to transform the environmental bioavailable iron into a complete and fully functional magnetic organelle. MB process involves different steps such as iron uptake, magnetosome vesicle formation, specific protein recruiting, crystal nucleation, redox balance, and pH control in magnetosome vesicle, size and crystalline morphology control and magnetosome vesicle docking in the bacterial cytoskeleton [3].

Mam and Mms proteins involved in MB belong to different protein families including: TPR proteins (from Tetratrico Peptide Repeat; MamA) [17], CDF transporters (Cation Diffusion Facilitators: MamB and MamM) [14, 18], serine proteases HtrA-like (MamE, MamP, and MamO) [14], actine-like proteins (MamK) [19], liposome tubulation protein (MamY) [20], generic transporters (MamH and MamN) [14, 21] and MTB specific proteins without prior homology in other non-magnetotactic microrganisms

(MamG, MamF, MamD, MamC, MamJ, MamW, MamX, MamY, Mms6, MtxA) [3]. MB involves four major steps as they are: (i) MM formation (participation of MamI, MamL and MamAB proteins) [3, 14]; (ii) crystal nucleation (which include MamE, Mms6, MamB and MamM) [3, 14]; (iii) crystal maturation (participation of MamE, MmsF, MamGFDC and Mam P, S, T) [3, 14]; and (iv) magnetosome chain alignment within cell body (participation of MamJ, MamK and MamY) [14, 20]. Mam and Mms protein functions involved in MB are described in **Table 1** and **Figure 3**.

Protein	*Operon*	Function	MTB strain	Reference
MamA	*mam*AB	Protein recruitment	AMB-1	[22]
MamB	*mam*AB	Membrane invagination and iron uptake	AMB-1/ MSR-1	[14, 18]
MamC	*mam*GFDC	Size and morphology control	AMB-1	[23]
MamD	*mam*GFDC	Size and morphology control	AMB-1	[23]
MamE	*mam*AB	Protein targeting and redox control	AMB-1	[14]
MamF	*mam*GFDC	Size control	AMB-1	[23]
MamG	*mam*GFDC	Size and morphology control	AMB-1	[23]
MamH	*mam*AB	Iron uptake	AMB-1/ MSR-1	[14, 21]
MamI	*mam*AB	Membrane invagination	AMB-1	[14]
MamJ	*mam*AB	Magnetosome alignment	MSR-1	[24]
MamK	*mam*AB	Magnetosome alignment	MSR-1	[19]
MamL	*mam*AB	Membrane invagination	AMB-1	[14]
MamM	*mam*AB	Iron uptake	AMB-1/ MSR-1	[14, 18]
MamN	*mam*AB	pH control	AMB-1	[14]
MamO	*mam*AB	Crystal nucleation	AMB-1/ MSR-1	[14, 25]
MamP	*mam*AB	Redox control	AMB-1	[14]
MamQ	*mam*AB	Membrane invagination	AMB-1	[14]
MamR	*mam*AB	Size and morphology control	AMB-1	[14]
MamS	*mam*AB	Size and morphology control	AMB-1	[14]
MamT	*mam*AB	Size e morphology control and redox control	AMB-1	[26]
MamU	*mam*AB	Not defined	AMB-1	[14]
MamV	*mam*AB	Not defined	MSR-1	[18]
MamW	*mam*AB	Magnetosome alignment	MSR-1	[27]
MamX	*mam*XY	Redox control	MSR-1	[21]
MamY	*mam*XY	Membrane invagination and magnetosome alignment	AMB-1/ MSR-1	[20, 28]
MamZ	*mam*XY	Iron uptake and redox control	MSR-1	[21]
Mms6	*mms6*	Size and morphology control	AMB-1	[29]
MmsF	*mms6*	Size and morphology control	AMB-1	[30]

Table 1.
Mam and Mms protein functions inferred by mutant construction in the cultured magnetotactic alphapreoteobacteria Ms. magneticum strain AMB-1 and Ms. gryphiswaldense strain MSR-1.

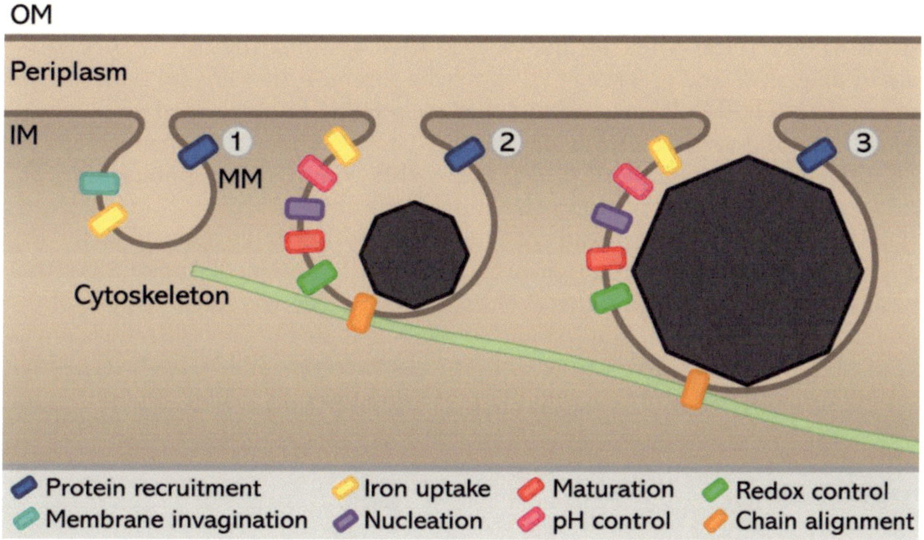

Figure 3.
Three major steps of MB in MTB. 1st step: protein recruitment initiating the biomineralization process while forming the invagination of the magnetosome membrane (MM) and iron uptake. 2nd step: Crystal nucleation, characterized by the incorporation of iron and oxygen for magnetite biomineralization. Interestingly, oxygen for the synthesis of magnetite is derived from water [31]. So far, the sulfur source for the synthesis of greigite has not been clarified. Magnetosome begins to grow in size while morphology, pH and redox balance are strictly regulated. Magnetosomes are aligned in chains within the cell's cytoskeleton. 3rd step: Magnetosomes continue to grow under strict regulation until de crystal maturation is complete. OM: outer membrane; IM: inner membrane, meaning the cytoplasmic membrane.

The advances of molecular biology techniques provided a much greater understanding of the MB mechanism over the last years as cultured and environmental MTB had their genomes sequenced. Magnetite MGCs and magnetite magnetosomes were studied in magnetotactic proteobacteria affiliated to the classes Alpha- [32–36], Beta- [37], Gamma- [38, 39], Delta- [40–43], *Ca.* Eta- [9, 32, 44, 45], *Ca.* Lambda- [9] and Zetaproteobacteria [9] and MTB affiliated to Nitrospirae [13, 16, 46–50] and Omnitrophica [49] phyla. Greigite MGC and greigite magnetosomes were characterized in magnetotactic deltaproteobacteria [51, 52] and MTB affiliated to *Ca.* Latescibacteria [8] and Planctomycetes [9] phyla. Culturing environmental MTB and mutant constructs different from the already known magnetotactic alphaproteobacteria *Ms. magneticum* strain AMB-1 and *Ms. gryphiswaldense* strain MSR-1 may provide a greater comprehension of the MB mechanism.

3. Evolutionary history of MGCs within Bacteria domain

MGC origin and evolution within the Bacteria domain is a constantly discussed topic in the literature. The scattering of MGCs and the magnetotactic behavior raises questions as MTB encompasses high diversity regarding their ecology, metabolism, and phylogeny. The first proposed hypothesis was the polyphyletic origin of magnetite and greigite MB [53]. According to this hypothesis, biomineralization of greigite and magnetite magnetosomes would have evolved without sharing a last universal common ancestor of magnetotactic bacteria (LUCA MTB). At that time MGCs were not discovered. Thus, this assumption relied on the information that the biochemical and nutritional parameters for greigite and magnetite biomineralization are different. Likewise, all known MTB affiliated to Alphaproteobacteria synthesized magnetite magnetosomes, while the ones affiliated to Deltaproteobacteria synthesized greigite magnetosomes, thus permitting the inference the polyphyletic hypothesis. Years later, after the discovery of MGCs, similarities between *mam* genes of magnetite and greigite MTB showed a common ancestor for both minerals synthesis in MTB [54]. It is speculated that greigite MGCs originated after events of duplication and divergence from magnetite MGCs in sulfate-reducing bacteria like the multicellular magnetotactic prokaryote (MMP) *Ca.* Magnetoglobus multicellularis strain Araruama affiliated to Deltaproteobacteria [54].

On behalf of that, Lefèvre and colleges [55] hypothesized a monophyletic origin of MGCs concerning magnetotactic proteobacteria. The comparison of 16S rRNA gene and conserved Mam proteins evolution showed a convergence of both phylogenetic inferences. It was suggested that MTB affiliated to Proteobacteria phyla shared a LUCA MTB and over time, some proteobacteria would have lost the MGC, resulting in the inability of biomineralizing magnetosomes [55].

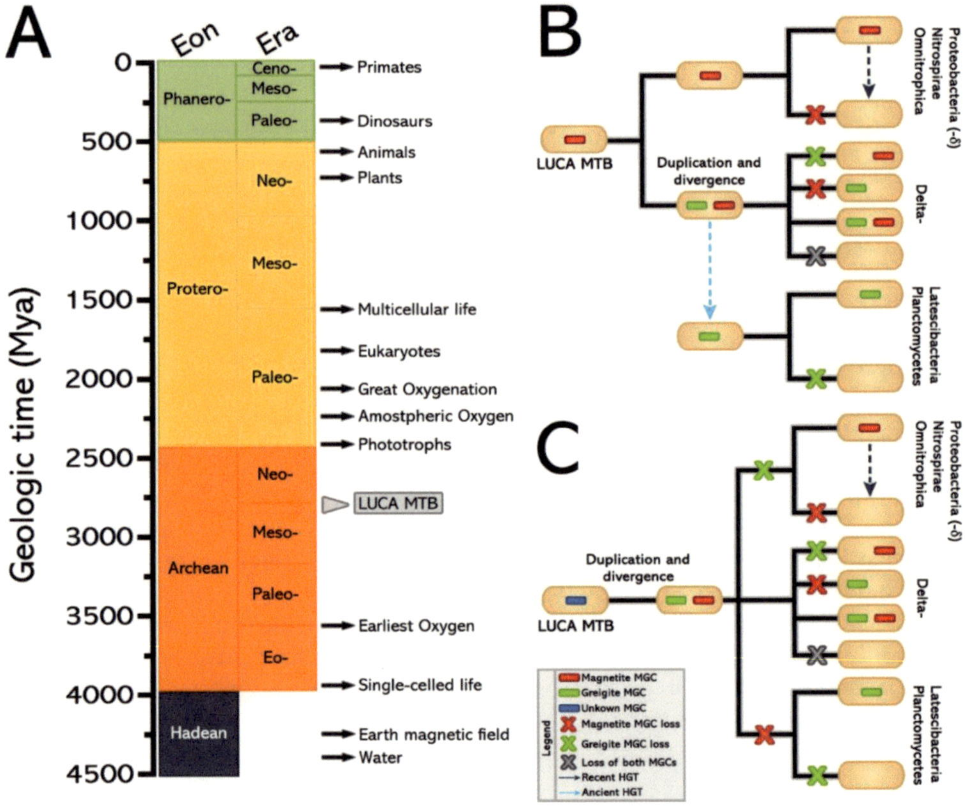

Figure 4.
Geologic time and evolution model proposed for MGC and magnetotaxis evolution. (A) Geologic rule in million years ago (Mya). LUCA MTB origin (gray arrowhead) is estimated 2.7 billion years ago during the Archean eon. The first single-celled form of life originated ~4 billion years ago and the origin of phototrophs, that permitted great oxygenation in earth, only happened ~2.4 billion years ago. (B and C) Two models for MGC and magnetotaxis evolution adapted from [9]. (B) LUCA MTB containing magnetite MGC branched two MTB lineages: (i) MTB affiliated to Proteobacteria (without Delta-), Nitrospirae and Omnitrophica phyla with recent HGT events responsible for MGC scattering; and (ii) MTB affiliated to Deltaproteobacteria class that after events of duplication and divergence hosted microbes with magnetite, greigite or both MGCs. Ancient HGT events would have been responsible for greigite MGC acquaintance in Plantomycetes and Ca. Lastescibacteria phyla. Adapted from [9]. (C) LUCA MTB containing an unknown MGC after events of duplication and divergence gave origin for both magnetite and greigite MGC. A monophyletic origin is proposed for MTB affiliated to Proteobacteria (without Delta- class), Nitrospirae, Omnitrophica, Planctomycetes and Ca. Latescibacteria phyla and Deltaproteobacteria class. Recent HGT events originating from MTB affiliated to Proteobacteria (without Delta- class), Nitrospirae, Omnitrophica could have been responsible for the scattering of MGC and magnetotactic behavior. Adapted from [9].

Opposing all previous statements, a considerable number of authors proposed the importance and influence of horizontal gene transfer (HGT) events on the evolution and scatter of MGC in Bacteria domain [9, 13, 56–59]. In light of these events, different non-MTB would have received MGCs by HGT, granting them the capacity of biomineralizing magnetosomes [9].

The origin of MB was dated, by molecular Bayesian clock, before the divergence of the Nitrospirae and Proteobacteria phyla during the Archean eon [13]. The divergence happened 2.7 billion years ago before the appearance of phototrophs and Great Oxygenation at the time of Paleoproterozoic on the Proterozoic eon (**Figure 4**). This hypothesis is supported by: (i) low pressure or absence of O_2 in the atmosphere and anoxic oceans in Archean [60]; (ii) abundant dissolved Fe^{2+} as concentrations of 40 to 120 μmol/L [61]; (iii) presence of primary electron donors of Earth early ecosystems such as H_2, H_2S, S^0, Fe^{2+}, CH_4, NH_4^+ and CH_2O [62]; and (iv) presence of primary electron acceptors of Earth early ecosystems such as CO_2, CO, SO_4^{2-}, NO, NO_2^- and NO_3^- [62]. These conditions favored the survival and growth of MTB [13]. Known examples of such conditions that are in accordance with available resources of primitive Earth are: (i) microaerophilic or anaerobic respiration in all known MTB; (ii) chemolithoautotrophy as MTB are capable of CO_2 fixation by Calvin–Benson–Bassham cycle, the reverse tricarboxylic acid cycle, or the reductive acetyl-CoA pathway [63]; (iii) capacity of denitrification of NO, NO_2^- and NO_3^- [16, 49]; (iv) capacity of oxidizing H_2S via sulfur oxidation pathway [16, 49]; (v) water temperature ranging from 26 to 85°C [64, 65] compatible with MTB growth as there are psychrophilic [66], mesophilic [8] and moderately thermophilic MTB [47]. Alongside these conditions, Earth's magnetic field originated 4.2 billion years ago enduring several inversions until the present time [67]. Considering this panorama, it is plausible that MTB and the geomagnetic fields have coevolved selecting the ones capable of undergoing all the continuous biotic and abiotic variations [13].

Large scale metagenome approach of MTB diversity demonstrated two possible routes concerning MGC evolution over time [9]. It is hypothesized that a LUCA MTB contained magnetite or an unknown MGC followed by events of MGC duplication, divergence, and loss combined with ancient and recent HGT events could explain the scattering of the magnetotactic behavior in the Bacteria domain [9] (**Figure 4**). The unending studies regarding MTB diversity and ecology are indispensable for an accurate decipherment of MGC evolution in the Bacteria domain.

4. Influence of the medium on biomineralization

The fact that related magnetotactic strains synthesize magnetosomes with significant differences in sizes and elongation is a clue that, despite a rigorous genetic control, environmental factors may influence the characteristics of the biomineralized nanocrystals [68]. Extensive experiments performed in cultures of MTB have pointed out temperature, pH, iron concentration, oxygen concentration, external magnetic fields, and nutrient concentrations as important factors driving physical changes in magnetosomes [69].

Ferric iron concentrations exert an important influence on the magnetic properties of *Magnetospirillum magnetotacticum* strain MS-1 cells due to alterations within biogenic magnetite [70]. The coercive force (H_C), probably the most important criterion in the selection of magnetic nanoparticles for technological applications, is significantly affected [70]. The H_C was increased from 216 Oe when cells were cultured at 12 μM Fe^{3+} to 238 Oe at 68 μM [70].

In another study, it was shown that reducing conditions leads to an increase in magnetosomes crystals of *Ms. magneticum* strain AMB-1 in culture [71]. An oxidoreduction potential of 0 mV (neutral condition) led to a crystal diameter of 31.5 ± 1.3 nm, which augmented to 37.2 ± 0.6 nm when the culture was carried out at -500 mV (reducing condition) [71]. The reducing condition also caused an increase in the total magnetite mass per cell as 9.1 ± 1.9 magnetosomes were observed per µm (cell length), in contrast to 5.48 ± 1.3 in neutral condition.

The evidence that characteristics of biogenic magnetite can be modified is of great interest for practical applications because certain purposes may require specific particle properties. Therefore, the knowledge of the interplay between environmental conditions and process regulation by biomolecules in biomineralization can help develop methods for the *in vitro* biomimetic preparation of magnetic nanoparticles with tunable properties.

5. Microbes inspire chemistry: biomimetic synthesis of artificial nanoparticles

Understanding MB is key not only for the in-depth learning of microbial physiological phenomena, but it can teach us valuable insights for the fabrication of technological materials. Magnetic nanoparticles have emerged as functional materials since the 1940s, when iron oxide powders, with crystals ranging from 60 nm to 1 µm, were used to impregnate recording tapes [72]. In that media, recorded information was engraved through changes in magnetization of the impregnated nanoparticles. Similarly, the biogenic magnetosomes can carry paleomagnetic signals, which can be detected, for instance, through the measurement of their magnetic properties in marine sediments [73]. The roles of bacterial magnetite as magnetofossils is only possible due to their stable single magnetic domain, caused by their controlled size range (20–100 nm) [73, 74]. This magnetic property also permits the utilization of biogenic nanomagnets in research on anticancer and antimicrobial therapy—as drug carriers, contrast agents, and hyperthermal agents—, enzyme immobilization—as recyclable supports—, cell labeling and other applications [11].

Biological materials are precisely arranged at the nanoscale. Hence, biomimetics, which is the art of imitating biological process to architecture novel materials, is proving profitable for nanotechnology industries [75]. One of the foundations of biomimetics is the biodiscovery and bioengineering of surface-binding proteins and peptides [76]. The regular structures present in such biomolecules enables the recognition and the interaction with atomic patterns on the surface of synthetic polymers, semiconductors, and metal oxide crystals [76]. In the case of metal oxides, these interactions occur basically via non-covalent weak bindings like hydrogen bonds and electrostatic dipoles.

In chemical syntheses, the shape- and size-controlled nanoparticles generally are obtained with high temperatures and organic solvents [74]. These consumptions are related to high production costs and environmental impacts during the life cycle of the nanoparticles [74]. One of the simplest and widely utilized techniques for making iron oxide nanoparticles is coprecipitation [74]. In this technique, ferrous and ferric salts are dissolved, and the cations are precipitated in an alkaline aqueous medium. For the synthesis of magnetite, a fixed molar proportion of 2:1 (Fe^{3+}/Fe^{2+}), is precipitated, following the stoichiometry:

$$2Fe^{3+} + Fe^{2+} + 8OH^{-} \rightarrow Fe_3O_4 + 4H_2O$$

This molar proportion is mandatory because it is the same ferrous/ferric ratio within magnetite [77]. In MTB, iron is accumulated inside the magnetosome vesicle in it ferrous form before being oxidized to ferric ion by magnetochromes—oxidizing domains of MamP, MamX, MamT and MamE [77]. This is an example of naturally occurring partial oxidation of ferrous ion. Partial oxidation is also used to obtain artificial, biomimetic magnetite [78]. In this case, the ferrous cation is precipitated to form ferrous hydroxide ($Fe(OH)_2$). After that, a strong oxidizing agent, usually nitrate, partially transform Fe^{2+} to Fe^{3+}, leading to magnetite:

$$Fe^{2+} + 2OH^{-} \rightarrow Fe(OH)_2$$

$$3Fe(OH)_2 + NO_3^{-} \rightarrow Fe_3O_4 + 3H_2O + NO_2^{-}$$

While coprecipitation leads to nanoparticles of an irregular shape, partial oxidation magnetite has a well-defined faceted morphology and a larger size [78]. Due to its low solubility, $Fe(OH)_2$ tends to form larger precipitates. This is not the case for the coprecipitation of Fe^{3+} and Fe^{2+}, which tends to form multiple, smaller precipitates [78].

Complementary to oxidation control, the surface interaction of the forming magnetic crystal with biomolecules is the main strategy for synthesizing magnetosome-like nanoparticles. A summary of biomolecule-supplemented chemical syntheses of magnetic nanoparticles is in **Table 2**.

MamC protein from *Magnetococcus marinus* strain MC-1 has an effect of enlarging magnetite precipitates [79, 84]. Due to its effect over synthesis, this protein has been expressed for use in different biomimetics studies (**Figure 5**). Different coprecipitation experiments have shown an increase from ~10-25 nm, in control synthesis, to ~30-40 nm, when recombinant MamC from strain MC-1 is added in concentrations over 10 μg/mL [79, 84].

In another study, *Ms. magneticum* strain AMB-1-derived Mms6 displays a negative effect on average particle size – 20 nm length down from 32 nm in the control experiment – in partial oxidation and coprecipitation-derived magnetite [80]. Instead, its addition to the reactional medium narrows size distribution regardless of the chemical route. The presence of recombinant Mms6 derived from strain AMB-1 imprints the cubo-octahedral morphology of the naturally occurring magnetosomes onto chemically precipitated crystals. From experiments using mutant clones of strain AMB-1, it has been demonstrated that the anionic residues Asp123, Glu124, and Glu125 effectively participate as key residues of Mms6 for defining crystal morphology are in the protein binding to magnetite [88]. The interactions between these C-terminal side-groups and the magnetite surface ultimately respond for the strong morphology and size controlling character of Mms6 either in biologic or biomimetic mineralization [89].

To modulate/improve magnetite chemical synthesis by the use MB proteins, magnetite-interacting components (MICs) of three magnetite-associated proteins (MamC, Mms6, and Mms7) have been subjected to NMR studies to investigate their affinity and binding to the ferrous ion during coprecipitation [81]. In all cases, it has been a clear role of aspartate and glutamate residues to the affinity to the cation [81]. The strong binding of ferrous cation to four anionic residues is related to confinement of iron by Mms6- and Mms7- MICs and, consequently, to the initiation of magnetite nucleation by these proteins. Besides ferrous ion, Mms6 glutamate residues positions 44, 50, and 55 at C-terminal region shows a strong binding affinity to ferric ion [90]. MamC-MIC, in turn, displays a weaker iron-binding but a stronger effect on magnetite size [81]. Thus, the ionotropic (i.e. iron-affinity) effect of MamC does not give sufficient ground for the role of this protein in

Additive(s)	Synthesis	Size (nm)	Shape	Ms (emu/g)	References
Magnetosomal proteins					
MamC	CP	30–40	Rhomboid	—	[79]
Mms6	PO	20.2 ± 4.0	Cubo-octahedral	—	[80]
MamC-MIC	CP	26.1 ± 0.61	Cuboid	—	[81]
Mms6-MIC		19.9 ± 0.36	Rhomboid	—	
Mms7-MIC		18.54 ± 0.29	Cuboid	—	
MmsF	CP	36	Rounded	129	[82]
MamF		25	Irregular	44	
Active loop of MmsF	CP	50 ± 13	Cuboidal	90	[83]
Active loop of Mms13		34 ± 12	Irregular	93	
MamC + Mms6	CP	30 ± 10	Rhomboid	—	[84]
Aminoacids					
Lysine (0.1 to 10 mM)	CP	21 ± 7 to 29 ± 7	Rhomboid	67 (for 10 mM Lys)	[85]
Arginine (0.1 to 10 mM)		16 ± 7 to 19 ± 6	—	36 (for 10 mM Arg)	
Polyaminoacids and polypeptides					
Polyarginine	CP	35 ± 5	Irregular	—	[86]
Polyaspartate	PO	7.6 ± 1.5	Rounded	78	[78]
14-mer peptide (magnetite-binding domain + ovarian cancer target) + ginger extract	CP	7.35 ± 3.7	Irregular	48.9	[87]

M_s = magnetization saturation at 300 K; CP = coprecipitation; PO = partial oxidation; MIC = magnetite-interacting component.

Table 2.
Summary of methods for chemical synthesis of biomimetic magnetic nanoparticles.

biomineralization [84, 91, 92]. MamC must exert a template effect in magnetite formation [84]. In the MM, MamC is constituted by two transmembrane domains connected by alpha-helical looping, which contacts the forming magnetite within the magnetosome vesicle lumen [92]. The distance between iron-interacting residues Glu66 and Asp70 of the alpha-helical looping matches the iron interatomic distance within the magnetite surface plane. The alpha-helical conformation of the MamC-MIC ensures the proper positioning of the points of interaction with iron [91]. The complementary roles of MamC and Mms6 can be combined in a biomimetic synthesis, yielding large magnetosomes (30 ± 10 nm) with well-defined crystal faces [84].

Other MM proteins are also good candidates for use in biomimetics. MamF controls the size monodispersity of nanocrystals. In aqueous solution, this protein forms a self-aggregative proteinosome of approximately 36 nm [82]. When used as an additive in coprecipitation, homogeneously sized nanocrystals are obtained. As in MamC, Mms13 and MmsF have their active loops located between the two transmembrane domains [83]. These active loops were expressed in a chimeric coiled-coil scaffold protein, which was called Mms13cc and MmsFcc. The MmsFcc construct regulated the cuboidal morphology of the produced nanocrystals.

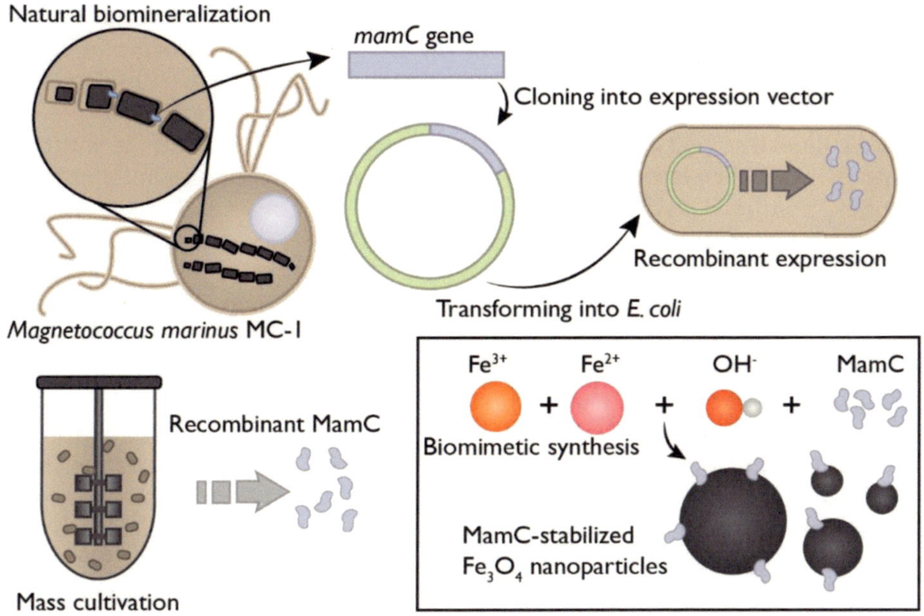

Figure 5.
Biomimetic route for making size- and shape-controlled magnetite nanoparticles [79]. MamC is a 12.4 kDa magnetite-interacting transmembrane protein found in different species of MTB. The gene encoding this protein in Mc. marinus strain MC-1 (mamC) was cloned in a pTrcHis-TOPO plasmidial expression vector. It was, then, transformed into Escherichia coli TOP10. The transforming E. coli can be cultivated in a large-volume (1–10 L) bioreactor and express the recombinant MamC. After mass-cultivation, expressed MamC, which is found in intracellular inclusion bodies, is recovered and then purified. MamC can be used as an additive for the coprecipitation of iron to synthesize nanometric magnetite. In this synthesis, MamC binds and stabilizes crescent magnetite nuclei. This interaction ultimately results in nanocrystals of narrowly distributed size and uniform morphology.

Taking the inspiration of the interaction between anionic residues and nascent magnetite, the addition of acidic polypeptides is an alternative to recombinant proteins [78]. In the presence of poly-aspartate, partial oxidation synthesis resulted in narrower size distribution of nanocrystals [78]. Using a classical partial oxidation synthesis, 65% of magnetite nanoparticles assumed a facetted shape with a size distribution between 20 and 60 nm. When the synthesis was supplemented with poly-aspartate, a drastic change of the morphology occurred, with 85% of the nanoparticles showing a more rounded shape. However, the size distribution became significantly narrower, with most particles ranging 15-30 nm.

As discussed, biomimetic synthesis of magnetite with recombinant magnetosome proteins involves electrostatic interaction between anionic aminoacids with iron cations. Nevertheless, the use of cationic polymers and aminoacids also has been proven successful in imitating characteristics of magnetosomes into artificial magnetite. In those cases, the one accepted chemical mechanism is the dipole stabilization of the negatively charged surface of magnetite crystals by positive side groups, namely amino and guanidine, present in alkaline aminoacids [85, 86]. This phenomenon is supported by the phosphatidylethanolamine composition of the magnetosome vesicle, which exposed positively charged amino groups to the nucleation sites [86, 93].

In one experiment performed at the Max Planck Institute of Colloids and Interfaces, Germany, a wide array of randomly-generated peptides was expressed in phage display and had their binding capacity tested against a magnetite powder [86]. The primary structure of magnetite adhering peptides was then compared to the proteomes of several MTB species, but no significant similarity was spotted.

However, of the five magnetite-interacting peptides identified in that study, three had arginine as half the residues in the sequence. The cationic poly-arginine was used as an additive to the iron precipitation. The resulting nanoparticles possessed a fine size distribution (30-40 nm), reproducible – despite irregular – morphologies and colloidal stability. These characteristics were not achieved in the control of conventional precipitation. Poly-arginine also improves the tuneability of the biomimetic synthesis. In the presence of the additive, the average diameters of the magnetite precipitates could be adjusted from 10 to 40 nm when the reaction occurred in pHs from 9 to 11, respectively [94].

As polyaminoacids, single aminoacids can promote control over magnetic nanoparticle syntheses [85]. When arginine and lysine were tested for that purpose, the latter was able to control the particle size according to its concentration (**Table 2**) [85]. The side-chain amino group in lysine can perform a steadier stabilization of the anionic oxyhydroxide precursor of magnetite. Then, further growth of lysine-stabilized nuclei enables a larger crystal size with a better-defined hexahedral shape. The control over size and shape also reflects in the magnetic properties of the nanomaterial. The obtained nanoparticles displayed a superparamagnetic behavior, with a large magnetic moment and magnetization saturation (67 emu/g).

Not only is the size dispersity and morphology better controlled in biomimetic synthesis, but the colloidal stability of bioinspired nanomagnets is generally improved. The magnetic core of bare nanomagnets exerts an attractive force, possibly leading to instability to the colloidal suspension [78, 85]. When peptides are added to the precipitation media, functional groups of the same charge become exposed on the nanoparticle surface and counterbalance the attractive force with electrostatic repulsion [78, 85]. Due to the interaction of cationic amino groups with magnetite, carboxyl groups become exposed during coprecipitation with lysine [85]. Thus, the zeta-potential of those nanoparticles was -31 mV at physiological pH, while the control nanoparticles showed a 0 value. The synthesis of magnetite supplemented with poly-aspartate led to nanoparticles with surface-exposed carboxyl groups [78]. Therefore, the measured zeta potential was approximately -30 mV. Because suspension stability in aqueous media is crucial for biomedical applications, the colloidal stability obtained in biomimetic nanoparticles is a fundamental property.

The knowledge gained from biomimetic approaches was used to construct a double-stimuli-responsive nanoformulation consisting of a nanomagnet bound to the antiproliferative drug oxaliplatin [95]. The nanocrystal was synthesized by co-precipitation of iron ions in the presence of recombinant MamC. The magnetite-oxaliplatin bond was stable at pH 7.2. In acidic pH, the release of oxaliplatin was triggered. This release was further boosted by the application of an alternating magnetic field and the cytotoxicity against colorectal cancer cells was improved [95]. The responsive to alternating magnetic fields also enables MamC-derived magnetic nanoparticles to be used in hyperthermia treatments [96]. A 25 mg/mL suspension of the biomimetic nanoparticles exposed to an alternating field of 226 Oe at a 280 kHz frequency can cause a temperature increase of 16.7 °C (specific absorption rate = 47 W/g).

Another functional magnetic nanoparticle was coprecipitated in the presence of a bifunctional polypeptide and ginger extract [87]. The fourteen-residue-long polypeptide was designed from two heptapeptides: a magnetite binding domain and a cell-targeting domain with specificity to ovarian carcinoma cells. The metal-reducing and chelating activity of the ginger extract leads to nanoparticles averaging 10 nm in length and 48.9 emu/g of magnetization saturation. When different cell lines – A2780 (ovarian carcinoma) and L929 (mouse fibroblast) – were treated with the functional nanoparticle, the first group exhibited a particle uptake almost 5 times more intense.

6. Conclusion

In this chapter, we have summarized how the basic-science knowledge gained through molecular biology, phylogenetics, and metagenomics of MTB can be translated into tools of technological interest. Although the authors had not the pretentiousness of gathering extensive information available on the topic, the chapter evidences how cross-disciplinary research is crucial for understanding and applying such a complex biological phenomenon. This is especially true in a field in which intriguing discoveries are made at a fast pace.

Acknowledgements

We thank Unidade de Microscopia Multiusuário Souto-Padrón & Lins (UniMicro, UFRJ, Brazil) for the use of their microscopy facility.

This research was funded by Brazilian agencies CNPq, CAPES and FAPERJ.

Conflict of interest

The authors declare no conflicts of interest.

Author details

Tarcisio Nascimento Correa, Igor Nunes Taveira, Rogerio Presciliano de Souza Filho and Fernanda de Avila Abreu*
Universidade Federal do Rio de Janeiro, Rio de Janeiro, Brazil

*Address all correspondence to: fernandaaabreu@micro.ufrj.br

IntechOpen

© 2020 The Author(s). Licensee IntechOpen. This chapter is distributed under the terms of the Creative Commons Attribution License (http://creativecommons.org/licenses/by/3.0), which permits unrestricted use, distribution, and reproduction in any medium, provided the original work is properly cited.

References

[1] Blakemore R. Magnetotactic bacteria. Science. 1975;**190**(4212):377-379. DOI: 10.1126/science.170679

[2] Bellini S. On a unique behavior of freshwater bacteria. Chinese Journal of Oceanology and Limnology. 2009;**27**(1):3-5. DOI: 10.1007/s00343-009-0003-5

[3] Uebe R, Schüler D. Magnetosome biogenesis in magnetotactic bacteria. Nature Reviews Microbiology. 2016;**14**(10):621-637. DOI: 10.1038/nrmicro.2016.99

[4] Lefèvre CT, Bazylinski DA. Ecology, diversity, and evolution of magnetotactic bacteria. Microbiology and Molecular Biology Reviews. 2013;**77**(3):497-526. DOI: 10.1128/MMBR.00021-13

[5] Amor M, Tharaud M, Gélabert A, Komeili A. Single-cell determination of iron content in magnetotactic bacteria: Implications for the iron biogeochemical cycle. Environmental Microbiology. 2020;**22**(3):823-831. DOI: 10.1111/1462-2920.14708

[6] Lin W, Bazylinski DA, Xiao T, Wu LF, Pan Y. Life with compass: Diversity and biogeography of magnetotactic bacteria. Environmental Microbiology. 2014;**16**(9):2646-2658. DOI: 10.1111/1462-2920.12313

[7] Martins JL, Silveira TS, Abreu F, Silva KT, da Silva-Neto ID, Lins U. Grazing protozoa and magnetosome dissolution in magnetotactic bacteria. Environmental Microbiology. 2007;**9**(11):2775-2781. DOI: 10.1111/j.1462-2920.2007.01389.x

[8] Lin W, Pan Y, Bazylinski DA. Diversity and ecology of and biomineralization by magnetotactic bacteria. Environmental Microbiology Reports. 2017;**9**(4):345-356. DOI: 10.1111/1758-2229.12550

[9] Lin W, Zhang W, Zhao X, Roberts AP, Paterson GA, Bazylinski DA, et al. Genomic expansion of magnetotactic bacteria reveals an early common origin of magnetotaxis with lineage-specific evolution. ISME Journal. 2018;**12**(6):1508-1519. DOI: 10.1038/s41396-018-0098-9

[10] Bazylinski D, Lefèvre C. Magnetotactic bacteria from extreme environments. Life. 2013;**3**(2):295-307. DOI: 10.3390/life3020295

[11] Vargas G, Cypriano J, Correa T, Leão P, Bazylinski D, Abreu F. Applications of magnetotactic bacteria, magnetosomes and magnetosome crystals in biotechnology and nanotechnology: Mini-review. Molecules. 2018;**23**(10):1-25. DOI: 10.3390/molecules23102438

[12] Bazylinski DA, Frankel RB. Magnetosome formation in prokaryotes. Nature Reviews Microbiology. 2004;**2**(3):217-230. DOI: 10.1038/nrmicro842

[13] Lin W, Paterson GA, Zhu Q, Wang Y, Kopylova E, Li Y, et al. Origin of microbial biomineralization and magnetotaxis during the Archean. Proceedings of the National Academy of Sciences of the United States of America. 2017;**114**(9):2171-2176. DOI: 10.1073/pnas.1614654114

[14] Murat D, Quinlan A, Vali H, Komeili A. Comprehensive genetic dissection of the magnetosome gene island reveals the step-wise assembly of a prokaryotic organelle. Proceedings of the National Academy of Sciences of the United States of America. 2010;**107**(12):5593-5598. DOI: 10.1073/pnas.0914439107

[15] Lefèvre CT, Trubitsyn D, Abreu F, Kolinko S, Jogler C, de Almeida LGP, et al. Comparative genomic analysis of magnetotactic bacteria from the Deltaproteobacteria provides new insights into magnetite and greigite magnetosome genes required for magnetotaxis. Environmental Microbiology. 2013;**15**(10):2712-2735. DOI: 10.1111/1462-2920.12128

[16] Lin W, Deng A, Wang Z, Li Y, Wen T, Wu LF, et al. Genomic insights into the uncultured genus "Candidatus Magnetobacterium" in the phylum Nitrospirae. ISME Journal. 2014;**8**(12):2463-2477. DOI: 10.1038/ismej.2014.94

[17] Zeytuni N, Ozyamak E, Ben-Harush K, Davidov G, Levin M, Gat Y, et al. Self-recognition mechanism of mamA, a magnetosome-associated TPR-containing protein. promotes complex assembly. Proceedings of the National Academy of Sciences of the United States of America. 2011;**108**(33):480-487. DOI: 10.1073/pnas.1103367108

[18] Uebe R, Junge K, Henn V, Poxleitner G, Katzmann E, Plitzko JM, et al. The cation diffusion facilitator proteins MamB and MamM of *Magnetospirillum gryphiswaldense* have distinct and complex functions, and are involved in magnetite biomineralization and magnetosome membrane assembly. Molecular Microbiology. 2011;**82**(4):818-835. DOI: 10.1111/j.1365-2958.2011.07863.x

[19] Komeili A, Li Z, Newman DK, Jensen GJ. Magnetosomes are cell membrane invaginations organized by the actin-like protein MamK. Science . 2006 Jan 13 [cited 2012 Nov 15];311(5758):242-5. DOI: 10.1126/science.1123231

[20] Toro-Nahuelpan M, Giacomelli G, Raschdorf O, Borg S, Plitzko JM, Bramkamp M, et al. MamY is a membrane-bound protein that aligns magnetosomes and the motility axis of helical magnetotactic bacteria. Nature Microbiology. 2019;**4**(11):1978-1989. DOI: 10.1038/s41564-019-0512-8

[21] Raschdorf O, Müller FD, Pósfai M, Plitzko JM, Schüler D. The magnetosome proteins MamX, MamZ and MamH are involved in redox control of magnetite biomineralization in *Magnetospirillum gryphiswaldense*. Molecular Microbiology. 2013;**89**(5):872-886. DOI: 10.1111/mmi.12317

[22] Komeili A, Vali H, Beveridge TJ, Newman DK. Magnetosome vesicles are present before magnetite formation, and MamA is required for their activation. Proceedings of the National Academy of Sciences of the United States of America. 2004;**101**(11):3839-3844. DOI: 10.1073/pnas.0400391101

[23] Scheffel A, Gärdes A, Grünberg K, Wanner G, Schüler D. The major magnetosome proteins MamGFDC are not essential for magnetite biomineralization in *Magnetospirillum gryphiswaldense* but regulate the size of magnetosome crystals. Journal of Bacteriology. 2008;**190**(1):377-386. DOI: 10.1128/JB.01371-07

[24] Scheffel A, Gruska M, Faivre D, Linaroudis A, Plitzko JM, Schüler D. An acidic protein aligns magnetosomes along a filamentous structure in magnetotactic bacteria. Nature. 2006;**440**(7080):110-114. DOI: 10.1038/nature04382

[25] Yang W, Li R, Peng T, Zhang Y, Jiang W, Li Y, et al. mamO and mamE genes are essential for magnetosome crystal biomineralization in *Magnetospirillum gryphiswaldense* MSR-1. Research in Microbiology. 2010;**161**(8):701-705. DOI: 10.1016/j.resmic.2010.07.002

[26] Siponen MI, Adryanczyk G, Ginet N, Arnoux P, Pignol D.

Magnetochrome: A c-type cytochrome domain specific to magnetotatic bacteria. Biochemical Society Transactions. 2012;**40**(6):1319-1323. DOI: 10.1042/BST20120104

[27] Lohsse A, Ullrich S, Katzmann E, Borg S, Wanner G, Richter M, et al. Functional analysis of the magnetosome island in *Magnetospirillum gryphiswaldense*: The mamAB operon is sufficient for magnetite biomineralization. PLoS One. 2011;**6**(10):e25561. DOI: 10.1371/journal.pone.0025561

[28] Tanaka M, Arakaki A, Matsunaga T. Identification and functional characterization of liposome tubulation protein from magnetotactic bacteria. Molecular Microbiology. 2010;**76**(2):480-488. DOI: 10.1111/j.1365-2958.2010.07117.x

[29] Tanaka M, Mazuyama E, Arakaki A, Matsunaga T. Mmm6 protein regulates crystal morphology during nano-sized magnetite biomineralization in vivo. Journal of Biological Chemistry. 2011;**286**(8):6386-6392. DOI: 10.1074/jbc.M110.183434

[30] Murat D, Falahati V, Bertinetti L, Csencsits R, Körnig A, Downing K, et al. The magnetosome membrane protein, MmsF, is a major regulator of magnetite biomineralization in *Magnetospirillum magneticum* AMB-1. Molecular Microbiology. 2012;**85**(4):684-699. DOI: 10.1111/j.1365-2958.2012.08132.x

[31] Mandernack KW, Bazylinski DA, Shanks WC, Bullen TD. Oxygen and iron isotope studies of magnetite produced by magnetotactic bacteria. Science. 1999;**285**(5435):1892-1896. DOI: 10.1126/science.285.5435.1892

[32] Bazylinski DA, Williams TJ, Lefèvre CT, Berg RJ, Zhang CL, Bowser SS, et al. *Magnetococcus marinus* gen. nov., sp. nov., a marine, magnetotactic bacterium that represents a novel lineage (Magnetococcaceae fam. nov., Magnetococcales ord. nov.) at the base of the Alphaproteobacteria. Internationl Journal of Systematic and Evolutionary Microbiology. 2013;**63**(3):801-808. DOI: 10.1099/ijs.0.038927-0

[33] Matsunaga T, Okamura Y, Fukuda Y, Wahyudi AT, Murase Y, Takeyama H. Complete genome sequence of the facultative anaerobic magnetotactic bacterium *Magnetospirillum* sp. strain AMB-1. DNA Research. 2005;**12**(3):157-166. DOI: 10.1093/dnares/dsi002

[34] Wang Y, Lin W, Pana Y. High diversity of magnetotactic Deltaproteobacteria in a freshwater Niche. Applied Environmental Microbiology. 2013;**79**(8):2813-2817. DOI: 10.1128/AEM.03635-12

[35] Ji B, Da ZS, Arnoux P, Rouy Z, Alberto F, Philippe N, et al. Comparative genomic analysis provides insights into the evolution and niche adaptation of marine *Magnetospira* sp. QH-2 strain. Environmental Microbiology. 2014;**16**(2):525-544. DOI: 10.1111/1462-2920.12180

[36] Monteil CL, Perrière G, Menguy N, Ginet N, Alonso B, Waisbord N, et al. Genomic study of a novel magnetotactic Alphaproteobacteria uncovers the multiple ancestry of magnetotaxis. Environmental Microbiology. 2018;**20**(12):4415-4430. DOI: 10.1111/1462-2920.14364

[37] Abreu F, Leão P, Vargas G, Cypriano J, Figueiredo V, Enrich-Prast A, et al. Culture-independent characterization of a novel magnetotactic member affiliated to the Beta class of the Proteobacteria phylum from an acidic lagoon. Environmental Microbiology. 2018;**20**(7):2615-2624. DOI: 10.1111/1462-2920.14286

[38] Lefèvre CT, Viloria N, Schmidt ML, Pósfai M, Frankel RB, Bazylinski DA.

Novel magnetite-producing magnetotactic bacteria belonging to the Gammaproteobacteria. ISME Journal. 2012;**6**(2):440-450. DOI: 10.1038/ismej.2011.97

[39] Leão P, Teixeira LCRS, Cypriano J, Farina M, Abreu F, Bazylinski DA, et al. North-seeking magnetotactic gammaproteobacteria in the southern hemisphere. Appl Environmental Microbiology. 2016;**82**(18):5595-5602. DOI: 10.1128/AEM.01545-16

[40] Nakazawa H, Arakaki A, Narita-Yamada S, Yashiro I, Jinno K, Aoki N, et al. Whole genome sequence of *Desulfovibrio magneticus* strain RS-1 revealed common gene clusters in magnetotactic bacteria. Genome Research. 2009;**19**(10):1801-1808. DOI: 10.1101/gr.088906.108

[41] Lefèvre CT, Frankel RB, Pósfai M, Prozorov T, Bazylinski DA. Isolation of obligately alkaliphilic magnetotactic bacteria from extremely alkaline environments. Environmental Microbiology. 2011;**13**(8):2342-2350. DOI: 10.1111/j.1462-2920.2011.02505.x

[42] Zhou K, Zhang WY, Yu-Zhang K, Pan HM, Da ZS, Zhang WJ, et al. A novel genus of multicellular magnetotactic prokaryotes from the Yellow Sea. Environmental Microbiology. 2012;**14**(2):405-413. DOI: 10.1111/j.1462-2920.2011.02590.x

[43] Abreu F, Morillo V, Nascimento FF, Werneck C, Cantão ME, Ciapina LP, et al. Deciphering unusual uncultured magnetotactic multicellular prokaryotes through genomics. ISME Journal. 2014;**8**(5):1055-1068. DOI: 10.1038/ismej.2013.203

[44] Lefèvre CT, Bernadac A, Yu-Zhang K, Pradel N, Wu LF. Isolation and characterization of a magnetotactic bacterial culture from the Mediterranean Sea. Environmental Microbiology. 2009;**11**(7):1646-1657. DOI: 10.1111/j.1462-2920.2009.01887.x

[45] Morillo V, Abreu F, Araujo AC, De Almeida LGP, Enrich-Prast A, Farina M, et al. Isolation, cultivation and genomic analysis of magnetosome biomineralization genes of a new genus of south-seeking magnetotactic cocci within the Alphaproteobacteria. Frontiers in Microbiology. 2014;**5**(72):1-12. DOI: 10.3389/fmicb.2014.00072

[46] Jogler C, Wanner G, Kolinko S, Niebler M, Amann R, Petersen N, et al. Conservation of proteobacterial magnetosome genes and structures in an uncultivated member of the deep-branching Nitrospira phylum. Proceedings of the National Academy of Sciences of the United States of America. 2011;**108**(3):1134-1139. DOI: 10.1073/pnas.1012694108

[47] Lefèvre CT, Abreu F, Schmidt ML, Lins U, Frankel RB, Hedlund BP, et al. Moderately thermophilic magnetotactic bacteria from hot springs in Nevada. Applied Environmental Microbiology. 2010;**76**(11):3740-3743. DOI: 10.1128/AEM.03018-09

[48] Lefèvre CT, Frankel RB, Abreu F, Lins U, Bazylinski DA. Culture-independent characterization of a novel, uncultivated magnetotactic member of the Nitrospirae phylum. Environmental Microbiology. 2011;**13**(2):538-549. DOI: 10.1111/j.1462-2920.2010.02361.x

[49] Kolinko S, Richter M, Glöckner FO, Brachmann A, Schüler D. Single-cell genomics of uncultivated deep-branching magnetotactic bacteria reveals a conserved set of magnetosome genes. Environmental Microbiology. 2016;**18**(1):21-37. DOI: 10.1111/1462-2920.12907

[50] Qian XX, Liu J, Menguy N, Li J, Alberto F, Teng Z, et al. Identification of novel species of marine magnetotactic bacteria affiliated with Nitrospirae phylum. Environmental Microbiology Reports. 2019;**11**(3):330-337. DOI: 10.1111/1758-2229.12755

[51] Lins U, Keim C, Evans F, Farina M, Buseck P. Magnetite (Fe_3O_4) and greigite (Fe_3S_4) crystals in multicellular magnetotactic prokaryotes. Geomicrobiology Journal. 2007;**24**(1):43-50. DOI: 10.1080/01490450601134317

[52] Lefèvre CT, Menguy N, Abreu F, Lins U, Pósfai M, Prozorov T, et al. A cultured greigite-producing magnetotactic bacterium in a novel group of sulfate-reducing bacteria. Science. 2011;**334**(6063):1720-1723. DOI: 10.1126/science.1212596

[53] DeLong EF, Frankel RB, Bazylinski DA. Multiple evolutionary origins of magnetotaxis in bacteria. Science. 1993;**259**(5096):803-806. DOI: 10.1126/science.259.5096.803

[54] Abreu F, Cantão ME, Nicolás MF, Barcellos FG, Morillo V, Almeida LGP, et al. Common ancestry of iron oxide- and iron-sulfide-based biomineralization in magnetotactic bacteria. ISME Journal. 2011;**5**(10):1634-1640. DOI: 10.1038/ismej.2011.35

[55] Lefèvre CT, Trubitsyn D, Abreu F, Kolinko S, de Almeida LGP, de Vasconcelos ATR, et al. Monophyletic origin of magnetotaxis and the first magnetosomes. Environmental Microbiology. 2013;**15**(8):2267-2274. DOI: 10.1111/1462-2920.12097

[56] Schübbe S, Kube M, Wawer C, Heyen U, Meyerdierks A, Madkour MH, et al. Characterization of a spontaneous nonmagnetic mutant of *Magnetospirillum gryphiswaldense* reveals a large deletion comprising a putative magnetosome island. Journal of Bacteriology. 2003;**185**(19):5779-5790. DOI: 10.1128/JB.185.19

[57] Richter M, Kube M, Bazylinski DA, Lombardot T, Glöckner FO, Reinhardt R, et al. Comparative genome analysis of four magnetotactic bacteria reveals a complex set of group-specific genes implicated in magnetosome biomineralization and function. Journal of Bacteriology. 2007;**189**(13):4899-4910. DOI: 10.1128/JB.00119-07

[58] Jogler C, Kube M, Schübbe S, Ullrich S, Teeling H, Bazylinski DA, et al. Comparative analysis of magnetosome gene clusters in magnetotactic bacteria provides further evidence for horizontal gene transfer. Environmental Microbiology. 2009;**11**(5):1267-1277. DOI: 10.1111/j.1462-2920.2009.01854.x

[59] Jogler C, Schüler D. Genomics, genetics, and cell biology of magnetosome formation. Annual Reviews of Microbiology. 2009;**63**:501-521. DOI: 10.1146/annurev.micro.62.081307.162908

[60] Canfield DE. The early history of atmospheric oxygen: Homage to Robert M. Garrels. Annual Reviews of Earth and Planetary Sciences. 2005;**33**:1-36. DOI: 10.1146/annurev.earth.33.092203.122711

[61] Rouxel OJ, Bekker A, Edwards KJ. Iron isotope constraints on the Archean and Paleoproterozoic Ocean redox state. Science. 2005;**307**(5712):1088-1091. DOI: 10.1126/science.1105692

[62] Canfield DE, Rosing MT, Bjerrum C. Early anaerobic metabolisms. Philosophical Transactions of the Royal Society B: Biological Sciences. 2006;**361**(1474):1819-1834. DOI: 10.1098/rstb.2006.1906

[63] Williams TJ, Zhang CL, Scott JH, Bazylinski DA. Evidence for autotrophy via the reverse tricarboxylic acid cycle in the marine magnetotactic coccus strain MC-1. Applied Environmental Microbiology. 2006;**72**(2):1322-1329. DOI: 10.1128/AEM.72.2.1322

[64] Blake RE, Chang SJ, Lepland A. Phosphate oxygen isotopic evidence for a temperate and biologically active Archaean Ocean. Nature. 2010;**464**(7291):1029-1032. DOI: 10.1038/nature08952

[65] Knauth LP. Temperature and salinity history of the Precambrian Ocean: Implications for the course of microbial evolution. Palaeogeography, Palaeoclimatology, Palaeoecology. 2005;**219**(1-2):53-69. DOI: 10.1016/j.palaeo.2004.10.014

[66] Abreu F, Carolina A, Araujo V, Leão P, Silva KT, de CFM, et al. Culture-independent characterization of novel psychrophilic magnetotactic cocci from Antarctic marine sediments. Environmental Microbiology. 2016;**18**(12):4426-4441. DOI: 10.1111/1462-2920.13388

[67] Tarduno JA, Cottrell RD, Davis WJ, Nimmo F, Bono RK. A hadean to Paleoarchean geodynamo recorded by single zircon crystals. Science. 2015;**349**(6247):521-524. DOI: 10.1126/science.aaa9114

[68] Pósfai M, Lefèvre CT, Trubitsyn D, Bazylinski DA, Frankel RB. Phylogenetic significance of composition and crystal morphology of magnetosome minerals. Frontiers in Microbiology. 2013;**4**:1-15. DOI: 10.3389/fmicb.2013.00344

[69] Moisescu C, Ardelean II, Benning LG. The effect and role of environmental conditions on magnetosome synthesis. Frontiers in Microbiology. 2014;**5**:1-12. DOI: 10.3389/fmicb.2014.00049

[70] Yiriletu IT. Magnetic properties of magnetite synthesized by *Magnetospirillum magnetotacticum* MS-1 cultured with different concentrations of ferric iron. Biotechnology Letters. 2015;**37**(12):2427-2433. DOI: 10.1007/s10529-015-1928-8

[71] Olszewska-Widdrat A, Schiro G, Reichel VE, Faivre D. Reducing conditions favor magnetosome production in magnetospirillum magneticum AMB-1. Frontiers in Microbiology. 2019;**10**(582):1-10. DOI: 10.3389/fmicb.2019.00582

[72] Mee CD. Magnetic tape recording materials. IEEE Transactions on Communication and Electronics. 1964;**83**(73):399-408. DOI: 10.1109/tcome.1964.6541245

[73] Amor M, Busigny V, Durand-Dubief M, Tharaud M, Ona-Nguema G, Gélabert A, et al. Chemical signature of magnetotactic bacteria. Proceedings of the National Academy of Sciences of the United States of America. 2015;**112**(6):1699-1703. DOI: 10.1073/pnas.1414112112

[74] Mirabello G, Lenders JJM, Sommerdijk NAJM. Bioinspired synthesis of magnetite nanoparticles. Chemical Society Reviews. 2016;**45**(18):5085-5106. DOI: 10.1039/C6CS00432F

[75] Bhushan B. Bioinspired materials and surfaces for green science and technology (part 3). Philosophical Transactions of the Royal Society A: Mathematical, Physical and Engineering Sciences. 2020;**378**(2167):2-3. DOI: 10.1098/rsta.2019.0439

[76] Sawada T, Serizawa T. Peptides as smart biomolecular tools: Utilization of their molecular recognition for materials engineering. In: Ito Y, Chen X, Kang I-K, editors. Advances in Bioinspired and Biomedical Materials. Washington, D.C.: American Chemical Society; 2017. pp. 31-48. DOI: 10.1021/bk-2017-1252.ch003

[77] Barber-Zucker S, Zarivach R. A look into the biochemistry of magnetosome biosynthesis in magnetotactic bacteria. ACS Chemical Biology. 2017;**12**(1):13-22. DOI: 10.1021/acschembio.6b01000

[78] Altan CL, Lenders JJM, Bomans PHH, De With G, Friedrich H, Bucak S, et al. Partial oxidation as a rational approach to kinetic control in bioinspired magnetite synthesis. Chemistry - A European Journal. 2015;**21**(16):6150-6156. DOI: 10.1002/chem.201405973

[79] Valverde-Tercedor C, Montalbán-López M, Perez-Gonzalez T, Sanchez-Quesada MS, Prozorov T, Pineda-Molina E, et al. Size control of in vitro synthesized magnetite crystals by the MamC protein of *Magnetococcus marinus* strain MC-1. Applied Microbiology and Biotechnology. 2015;**99**(12):5109-5121. DOI: 10.1007/s00253-014-6326-y

[80] Amemiya Y, Arakaki A, Staniland SS, Tanaka T, Matsunaga T. Controlled formation of magnetite crystal by partial oxidation of ferrous hydroxide in the presence of recombinant magnetotactic bacterial protein Mms6. Biomaterials. 2007;**28**(35):5381-5389. DOI: 10.1016/j.biomaterials.2007.07.051

[81] Nudelman H, Lee YZ, Hung YL, Kolusheva S, Upcher A, Chen YC, et al. Understanding the biomineralization role of magnetite-interacting components (MICs) from magnetotactic bacteria. Frontiers in Microbiology. 2018;**9**(2480):1-14. DOI: 10.3389/fmicb.2018.02480

[82] Rawlings AE, Bramble JP, Walker R, Bain J, Galloway JM, Staniland SS. Self-assembled MmsF proteinosomes control magnetite nanoparticle formation in vitro. Proceedings of the National Academy of Sciences of the United States of America. 2014;**111**(45):19094-19099. DOI: 10.1073/pnas.1409256111

[83] Rawlings AE, Somner LA, Fitzpatrick-Milton M, Roebuck TP, Gwyn C, Liravi P, et al. Artificial coiled coil biomineralisation protein for the synthesis of magnetic nanoparticles. Nature Communications. 2019;**10**(1):1-9. DOI: 10.1038/s41467-019-10578-2

[84] Peigneux A, Jabalera Y, Vivas MAF, Casares S, Azuaga AI, Jimenez-Lopez C. Tuning properties of biomimetic magnetic nanoparticles by combining magnetosome associated proteins. Scientific Reports. 2019;**9**(1):1-11. DOI: 10.1038/s41598-019-45219-7

[85] Contreras-Montoya R, Jabalera Y, Blanco V, Cuerva JM, Jimenez-Lopez C, Alvarez de Cienfuegos L. Lysine as size-control additive in a bioinspired synthesis of pure superparamagnetic magnetite nanoparticles. Crystal Growth & Design. 2020;**20**(2):533-542. DOI: 10.1021/acs.cgd.9b00169

[86] Baumgartner J, Antonietta Carillo M, Eckes KM, Werner P, Faivre D. Biomimetic magnetite formation: From biocombinatorial approaches to mineralization effects. Langmuir. 2014;**30**(8):2129-2136. DOI: 10.1021/la404290c

[87] Liu L, Pu X, Yin G, Chen X, Yin J, Wu Y. Biomimetic mineralization of magnetic iron oxide nanoparticles mediated by bi-functional copolypeptides. Molecules. 2019;**24**(7):16. DOI: 10.3390/molecules24071401

[88] Yamagishi A, Narumiya K, Tanaka M, Matsunaga T, Arakaki A. Core amino acid residues in the morphology-regulating protein, Mms6, for intracellular magnetite biomineralization. Scientific Reports. 2016;**6**(35670):1-10. DOI: 10.1038/srep35670

[89] Yamagishi A, Tanaka M, Lenders JJM, Thiesbrummel J, Sommerdijk NAJM, Matsunaga T,

et al. Control of magnetite nanocrystal morphology in magnetotactic bacteria by regulation of mms7 gene expression. Scientific Reports. 2016;**6**:1-11. DOI: 10.1038/srep29785

[90] Rawlings AE, Liravi P, Corbett S, Holehouse AS, Staniland SS. Investigating the ferric ion binding site of magnetite biomineralisation protein Mms6. PLoS One. 2020;**15**(2):1-16. DOI: 10.1371/journal.pone.0228708

[91] Nudelman H, Perez Gonzalez T, Kolushiva S, Widdrat M, Reichel V, Peigneux A, et al. The importance of the helical structure of a MamC-derived magnetite-interacting peptide for its function in magnetite formation. Acta Crystallographica Section D: Structural Biology. 2018;**74**:10-20. DOI: 10.1107/S2059798317017491

[92] Ubago-Rodríguez A, Casares Atienza S, Fernández-Vivas A, Peigneux A, Jabalera Y, De La Cuesta-Rivero M, et al. Structure-function of MamC loop and its effect on the in vitro precipitation of biomimetic magnetite nanoparticles. Crystal Growth & Design. 2019;**19**(5):2927-2935. DOI: 10.1021/acs.cgd.9b00150

[93] Grünberg K, Müller E-C, Otto A, Reszka R, Linder D, Kube M, et al. Biochemical and proteomic analysis of the magnetosome membrane in *Magnetospirillum gryphiswaldense*. Applied Environmental Microbiology. 2004;**70**(2):1040-1050. DOI: 10.1128/AEM.70.2.1040-1050.2004

[94] Kuhrts L, Macías-Sánchez E, Tarakina NV, Hirt AM, Faivre D. Shaping magnetite with poly-l-arginine and pH: From small single crystals to large mesocrystals. Journal of Physical Chemistry Letters. 2019;**10**(18):5514-5518. DOI: 10.1021/acs.jpclett.9b01771

[95] Jabalera Y, Garcia-Pinel B, Ortiz R, Iglesias G, Cabeza L, Prados J, et al. Oxaliplatin–biomimetic magnetic nanoparticle assemblies for colon cancer-targeted chemotherapy: An in vitro study. Pharmaceutics. 2019;**11**(8):4-6. DOI: 10.3390/pharmaceutics11080395

[96] Iglesias GR, Jabalera Y, Peigneux A, Fernández BLC, Delgado ÁV, Jimenez-Lopez C. Enhancement of magnetic hyperthermia by mixing synthetic inorganic and biomimetic magnetic nanoparticles. Pharmaceutics. 2019;**11**(6):16. DOI: 10.3390/pharmaceutics11060273

Chapter 3

Nanostructuring Bi_2Te_3-Based Thermoelectric Thin-Films Grown Using Pulsed Laser Deposition

Le Thi Cam Tuyen, Phuoc Huu Le and Sheng-Rui Jian

Abstract

This book chapter reports recent advances in nanostructured Bi_2Te_3-based thermoelectric (TE) thin-films fabricated by pulsed laser deposition (PLD). By controlling the processing conditions in PLD growths, various fascinating Bi_2Te_3-based nanostructured films with promising or enhanced TE properties have been successfully fabricated, including super-assembling of Bi_2Te_3 hierarchical nanostructures, self-assembled Bi_2Te_3 films with well-aligned 0D to 3D nanoblocks, polycrystalline-nanostructured Bi_2Se_3 and Bi_2Te_3 thin-films, etc. In addition, a PLD-growth mechanism for fabricating the super-assembling Bi_2Te_3 thin-films is presented. This book chapter provides fundamental understanding the relationship amongst processing condition, structure-morphology, and TE property of PLD-growths Bi_2Te_3-based thin-films. It also presents an overview of TE materials and applications with the challenges and perspectives.

Keywords: Bi_2Te_3, thermoelectrics, self-assembly nanostructures, thermoelectric power factor, pulsed laser deposition

1. Introduction

Thermoelectric materials are solid-state energy converters whose combination of thermal, electrical, and semiconducting properties allows them to be used to convert waste heat into electricity or electrical power directly into cooling and heating [1].

1.1 Thermoelectric effects

When an electric current flows through a pair of p-type and n-type semiconductors connected in series (**Figure 1(a)**, the holes in the p-type material and the electrons in the n-type material carry heat away from the top metal–semiconductor junctions, which leads to a cooling at the junctions called the Peltier effect. When current flows within the module, one side is cooled and the other heated. If the current is reversed, the hot and cold sides reverse also. For each material, the cooling effect is gauged by the Peltier coefficient Π that relates the heat carried by the charges to the electrical current through [1, 2, 4]: $Q = \Pi \times I$.

In **Figure 1(b)**, when the two ends of the materials maintain a temperature difference, the higher thermal energy holes and electrons will diffuse from the hot

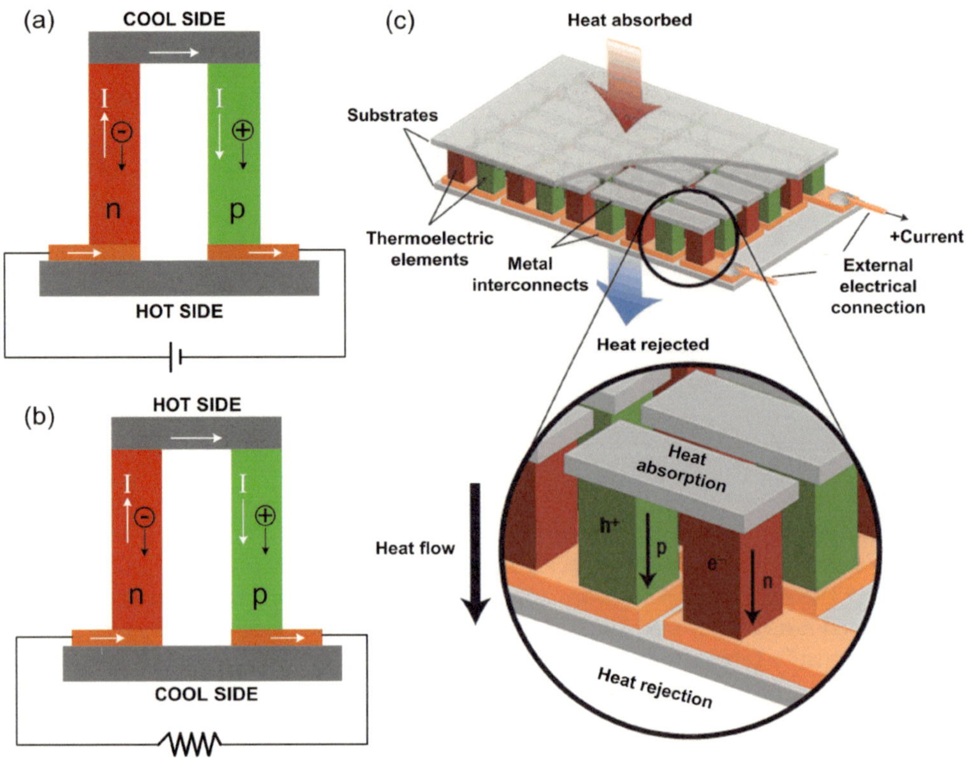

Figure 1.
Illustration of TE devices: (a) cooler (Peltier effect), (b) power generator (Seebeck effect). Redrawn after Ref. [2]. (c) Thermoelectric module showing the direction of charge flow on both cooling and power generation [3].

side to the cold side, and consequently a potential difference is created. This is Seebeck effect and it is the principle for thermocouples. The power generation is measured by the Seebeck coefficient α, which relates the voltage generated to the temperature difference through $\Delta V = -\alpha\Delta T$. The Peltier and the Seebeck coefficients are related through the Kelvin relation [1, 2]: $\Pi = \alpha T$.

Thermoelectric devices contain many thermoelectric couples (**Figure 1c**, bottom), which consist of p-type (containing free holes) and n-type (containing free electrons) thermoelectric elements connected electrically in series and thermally in parallel (**Figure 1c**, top). A thermoelectric generator uses heat flow across a temperature gradient to power an electric load through the external circuit.

1.2 The thermoelectric figure of merit (ZT)

The performance of the thermoelectric materials is often denoted as figure of merit Z whose unit is K^{-1}, or ZT the dimensionless unit [5, 6].

$$ZT = \frac{\alpha^2\sigma}{\kappa}T = \frac{\alpha^2\sigma}{\kappa_E + \kappa_L}T \tag{1}$$

where α, σ, T, and κ are the Seebeck coefficient, electrical conductivity, absolute temperature, and thermal conductivity, respectively. The total thermal conductivity can be split into electronic contribution (κ_E) and lattice contribution (κ_L). The thermoelectric power factor (PF) is calculated by the quantity of $\alpha^2\sigma$. The efficiency of a thermoelectric material is determined by its ZT. Meanwhile, the maximum efficiency (η) of a power generation is expressed by [3, 7]:

$$\eta = \frac{T_h - T_c}{T_h} \cdot \frac{\sqrt{1+Z\overline{T}} - 1}{\sqrt{1+Z\overline{T}} + \frac{T_c}{T_h}} \tag{2}$$

and the coefficient of performance presents for the efficiency of air-conditioning and refrigeration [7]:

$$COP = \frac{T_c}{T_h - T_c} \cdot \frac{\sqrt{1+Z\overline{T}} - \frac{T_h}{T_c}}{\sqrt{1+Z\overline{T}} + 1} \tag{3}$$

where T_h and T_c are the hot-end and cold-end temperature of the thermoelectric materials, respectively, and $\overline{T}$ is the average temperature of T_h and T_c. For practical applications, it is important to use high ZT thermoelectric materials.

The best materials so far are alloys of Bi_2Te_3 with Sb_2Te_3 and Bi_2Te_3 with Bi_2Se_3. ZT is of the order of 1 at room temperature. This value gives a COP of about 1 (**Figure 2a**), which is still far lower than the COP = 2–4 of household refrigerators and air conditioners. Similar situation is true for power generation (**Figure 2b**) [2, 8]. Thermoelectric cooling and power generation generally still not competitive with the other energy conversion methods.

1.3 Challenges in enhancing ZT

A concept of "phonon-glass electron-crystal" (or PGEC in short) was proposed for designing efficient thermoelectric materials. This is a controversial concept from the aspect of materials science that the materials should have a high electrical conductivity as in a crystal and a low lattice thermal conductivity as in a glass [9]. However, the TE parameters are strongly interdependent, which makes the enhancement efforts of ZT very challenging. A normal approach for the enhanced properties of TE materials is to increase the power factor $\alpha^2\sigma$ by optimizing the carrier concentration n, and/or to reduce the lattice thermal conductivity κ_L by introducing the scattering centers. These parameters are the function of carrier effective mass m^* and carrier mobility μ, scattering factor r, and their interconnectivity limit ZT to approximately 1 in large bulk materials [10].

The kinetic definition of α is the energy difference between the average energy of mobile carriers and the Fermi energy [11]. When carrier concentration (n) is increased, both the Fermi energy and the average energy increase, but the Fermi energy increases more rapidly than the average energy as n is increased. Consequently, α decreases and thus $\alpha^2 n$ is dragged down rapidly. Therefore, the carrier

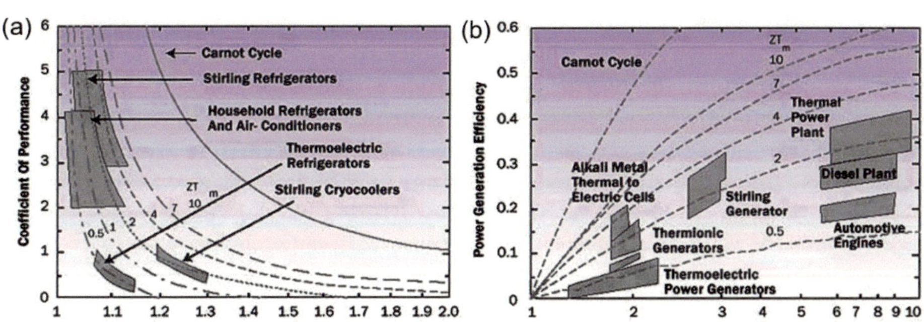

Figure 2.
Comparison of thermoelectric technology with other energy conversion methods for (a) cooling and (b) power generation [2, 8].

concentration (n) increases electrical conductivity (σ) but reduces the Seebeck coefficient (α) for most of the homogeneous materials. For this reason, in metals and degenerate semiconductors with energy-independent scattering approximation, the Seebeck coefficient can be expressed as [3, 12]:

$$\alpha = \frac{8\pi^2 k_B^2}{3eh^2} m^* T \left(\frac{\pi}{3n}\right)^{2/3} \tag{4}$$

Where, the parameter m^* is density of states effective mass, and an increase of m^* can raise the Seebeck coefficient according to the Eq. (4). However, most high m^* materials have generally low μ which limits the α by a weighted mobility with a factor proportional to $(m^*)^{3/2}\mu$. Moreover, there is no such thing as an optimal effective mass. There are high mobility low effective mass semiconductors (SiGe, GaAs) as well as low mobility high effective mass polaron conductors (oxides, chalcogenides) [3].

Noticeably, the defects scatter not only the phonons but also the electrons. When a thermoelectric material is designed for reducing lattice thermal conductivity, its carrier mobility is usually suppressed. Hence, the ratio of μ/κ_L determines the improvement of ZT [5, 10]. The ratio is observed to increase experimentally through a more reduction in κ_L rather than that in μ, but some fundamental issues in this mechanism are not understood well [10].

The electrical resistivity (ρ) and electrical conductivity (σ) are related to n through the carrier mobility μ:

$$1/\rho = \sigma = ne\mu \tag{5}$$

The electronic contribution to the thermal conductivity is proportional to the electrical conductivity (σ) of the materials according to Wiedemann–Franz Law [3], and the relationship is expressed as follows:

$$\kappa e = L\sigma T = ne\mu LT \tag{6}$$

where 'e' is electron charge, and L is Lorenz factor $2.48 \times 10^{-8}\ J^2/K^2C^2$ for free electrons and this can vary particularly with carrier concentration [3, 13].

Figure 3 shows the compromise of σ, κ and α in thermoelectric materials that must be optimized to maximize the figure of merit ZT. Indeed, the lower carrier concentration will result in the lower σ and a decreasing ZT. Typically, the PF and ZT peaks occur at carrier concentrations of 10^{19}–10^{21} cm^{-3} (depending on the material system), which falls in between common metals and heavily doped semiconductors [3]. High mobility carriers are most important for high value of electrical conductivity. Again from the Eq. (4), an increase of the carrier effective mass lead to increase the α but reduce the μ and hence the σ according to the Eq. (5). In case of the narrow semiconductor, the thermal excitation of carrier from valence band to conduction band creates holes and electrons. However, the concentration of the major carrier does not vary much. When two types of carriers are present, or bipolar effects takes place, and this is notorious to achieve effective thermoelectrics [4]. For example, the Seebeck coefficient for different carrier types is given by a weighted average of their electrical conductivity values (σ_e and σ_p) [13].

$$\alpha \approx \frac{\alpha_e\sigma_e + \alpha_p\sigma_p}{(\sigma_e + \sigma_p)} \tag{7}$$

In short, any attempt to increase σ, will increase κ_e which contributes to thermal conductivity (κ). In order to counter the increment of κ_e, various approaches are

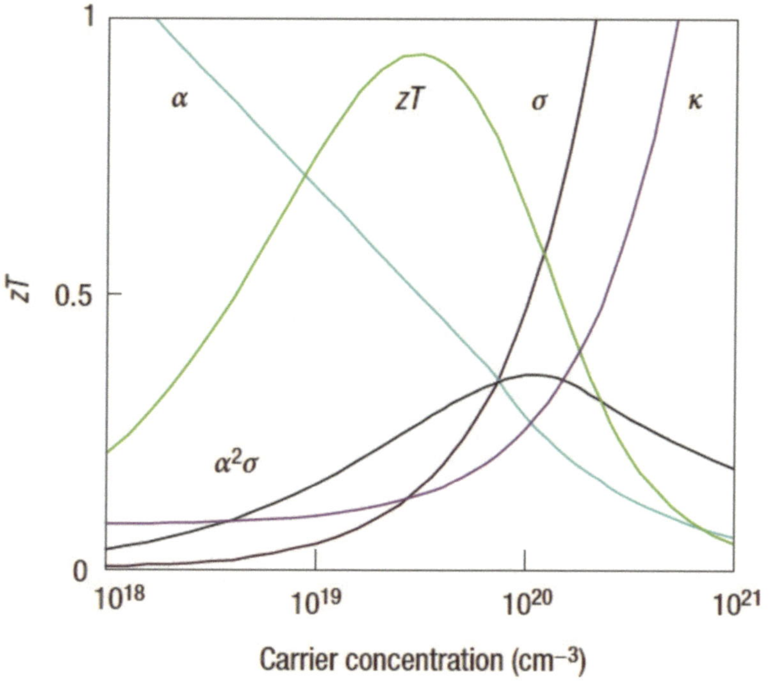

Figure 3.
Maximizing the efficiency (ZT) of a thermoelectric involves a compromise of thermal conductivity (κ; plotted on the y-axis from 0 to a top value of 10 Wm^{-1} K^{-1}) and Seebeck coefficient (α; 0–500 μVK^{-1}) with electrical conductivity (σ; 0–5000 Ω^{-1} cm^{-1}) [3].

reported to reduce κ_L. However, decreasing κ_L with phonon scattering by adding defects results in decrease in n and σ. These are the major conflicts in the properties of bulk thermoelectric materials which have been addressed in the researches for more than a half century [10].

1.4 Nanostructuring thermoelectric materials

In classical physics, the coefficients α, κ_e and σ are interrelated in such a way that it is impossible to increase one without affecting the others. Therefore, a compromise has to be achieved to find the maximum ZT value. Three different strategies have appeared to improve the ZT [14]:

a. An approach for increasing α while keeping the values of σ and κ_e by looking for new materials with complex band structures, like heavy fermion compounds.

b. Controlling the disorder in materials (such as Skutterudites or Clathrates) to present a rattling effect which causes, (↑) σ and decreases (↓) κ_L (see for instance ref. [15]).

c. Developing nanostructured materials that could lead to (↑) α due to quantum confinement effects, while ↓ κ_L due to the scattering of phonons at the interfaces. The latest improvements in the ZT of different materials has been achieved by this approach.

In 1993, Hicks and Dresselhaus pioneered the concept of nanostructuring in design of thermoelectric materials (i.e. Bi_2Te_3). The addition of the dimensionality and size of the system is added as a new parameter that affects the coupling of the electrical conductivity, Seebeck coefficient, and thermal conductivity, leading to substantially enhanced ZT [16–18]. Two ideas are dominant for the low-dimensional materials approach for improving ZT. Firstly, the presence of nano-scale constituents would introduce quantum confinement effects to enhance Seebeck coefficient and the power factor $\alpha^2\sigma$. Secondly, the numerous internal nanoinclusions and interfaces found in nanostructures would be designed so that the thermal conductivity would be reduced more than the electrical conductivity, based on differences in their respective scattering lengths [16].

As the dimensionality is decreased from 3D crystalline solids to 2D (quantum wells) to 1D (quantum wires) and finally to 0D (quantum dots), the spatial confinement are introduced that create the possibilities to tune the TE properties α, σ, and κ independently. When the system size decreases and approaches the scale comparable to the feature length of electron behavior (e.q. mean free path and wavelength) in any direction, the electronic density of states (D.O.S.) can split and become narrow as well as increase substantially (**Figure 4a**), resulting in the enhancement of α. Meanwhile, the thermal conductivity is also reduced because of the extensive phonon scattering at the surface, interfaces, and grain boundaries, as any dimension is less than the mean free path of phonons. **Figure 4(b)** illustrates

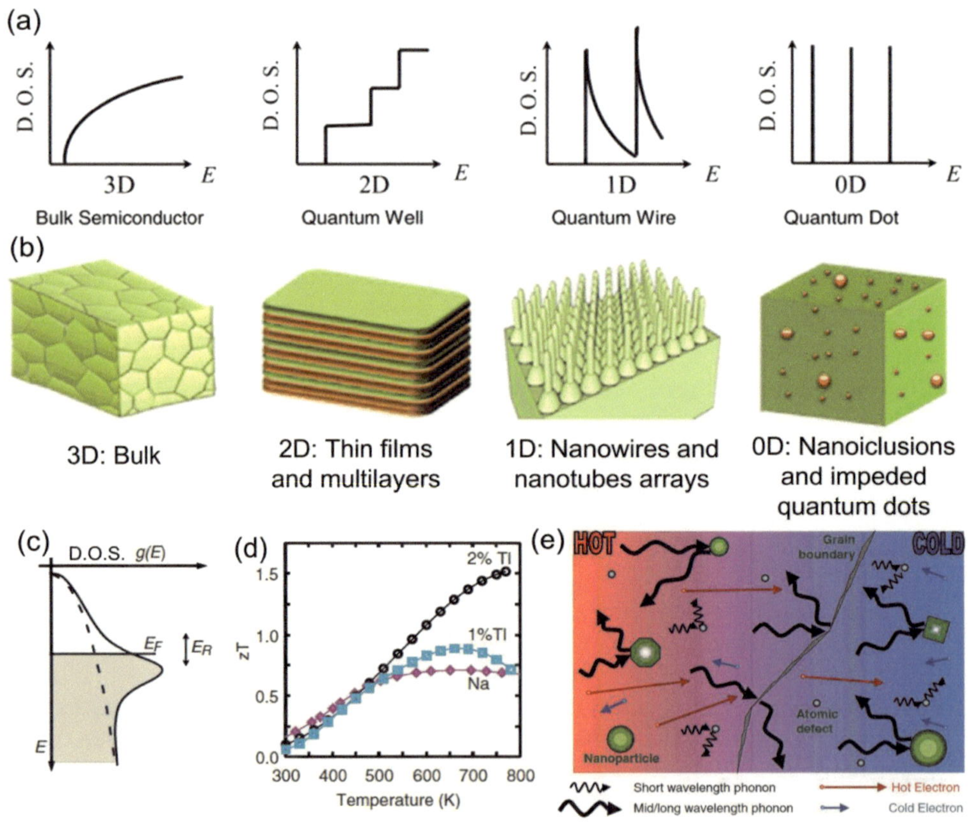

Figure 4.
(a) Electronic density of states (D.O.S.) for a bulk 3D crystalline semiconductor, a 2D quantum well, a 1D nanowire or nanotube, and a 0D quantum dot [16]. (b) Examples of different nanostructuring with different dimensionalities [14]. (c) A spike in the density of states (solid line) above the bulk value (dashed line) occurs due to resonant states in Tl-doped PbTe [19]. (d) The measured ZT of Tl-PbTe and Na-PbTe samples for 300–800 K indicates an improvement due to the addition of Tl [19]. (e) Schematic diagram illustrating various phonon scattering mechanisms within a thermoelectric material, along with electronic transport of hot and cold electrons [20].

examples of different nanostructuring with different dimensionalities [14]. A schematic diagram is shown in **Figure 4(e)** capturing these various phonon scattering mechanisms, along with the electrical transport within a thermoelectric material. For example, in material embedded nano-inclusions (nanoparticles), atomic defects are effective at scattering short wavelength phonons, but larger embedded nanoparticles are required to scatter mid- and long-wavelength phonons effectively. Grain boundaries can also play an effective role in scattering these longer-wavelength phonons [20].

Figure 5 plots major milestones achieved for ZT over the past several decades as a function of both year and temperature [20]. In the 1950s, Bi_2Te_3 was first investigated as a material of great thermoelectric with ZT~0.6 near room temperature [5, 6]. It was quickly realized that alloying with Sb_2Te_3 and Bi_2Se_3 allowed for the fine tuning of the carrier concentration alongside a reduction in lattice thermal conductivity. These compounds have played a dominant role in the field of thermoelectrics through today. The alloys of Bi_2Te_3 with Sb_2Te_3 (such as $Bi_{0.5}Sb_{1.5}Te_3$; p type) and of Bi_2Te_3 with Bi_2Se_3 (such as $Bi_2Te_{2.7}Se_{0.3}$; n type), with a ZT $\sim$ 1 at room temperature are traditional cooling materials [6]. In recent year, great enhancements in ZT owning to low dimension and nanostructure materials have been reported [19–32] and achieved the highest ZT value of approximately 2.4.

1.5 Overview of thermoelectric applications

The solid-state devices based on TE effect have the inherent advantages of reliability, silent and vibration-free operation (no moving fluids or moving parts), a very high power density, and the ability to maintain their efficiency in small scale applications where only a moderate amount of power is needed [19].

Commercial use has been made mostly from Peltier's thermoelectric cooling (TEC) effect in applications, as demonstrated in **Figure 6** [35]:

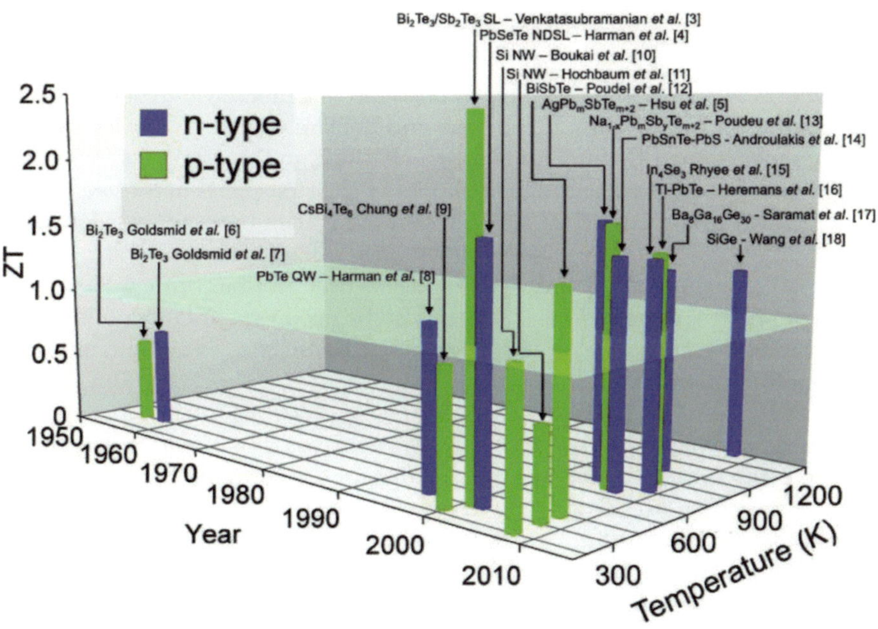

Figure 5.
Thermoelectric figure-of-merit ZT as a function of temperature and year illustrating important milestones [20]. Although there have been several demonstrations of ZT > 1 in the past decade (2001–2010), no material has yet achieved the target goal of ZT ≥ 3. The material systems that have achieved ZT > 1 have all been based on some form of nanostructuring.

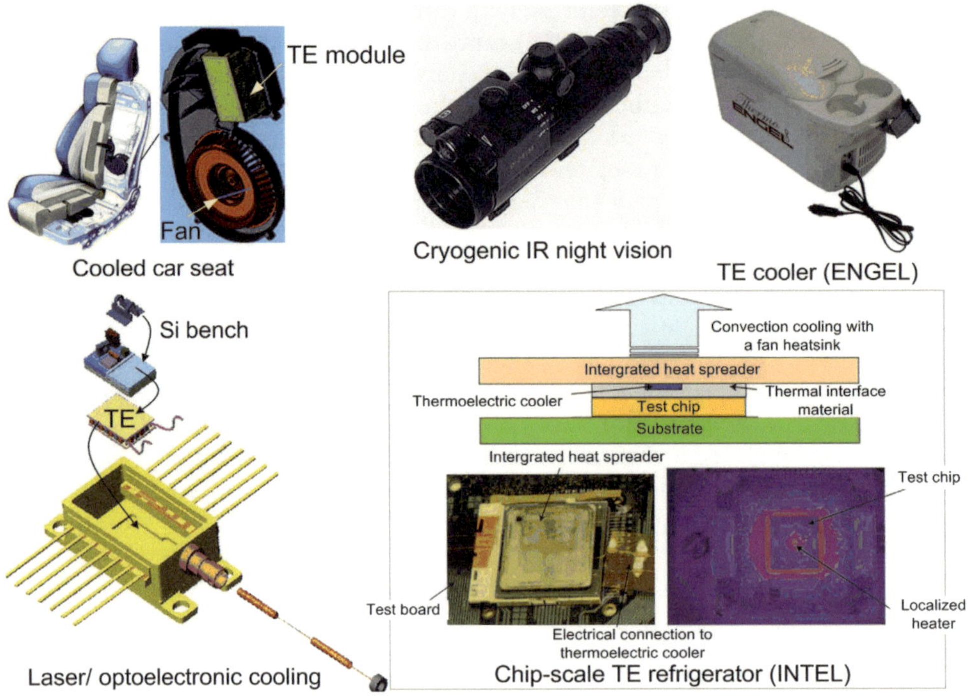

Figure 6.
Overview of potential thermoelectric cooling (TEC) applications [33, 34].

- Small refrigerator devices are used for camping and outdoor activities. For example, the cooler/warmer TE device (Engel Thermo 8) has volume 8 L and weighing just over 3 kg. Its features include cooling performance up to 22°C below ambient temperature and warming up to +65°C.

- Gentherm designed and developed Automotive Climate Control Seat [36], which has TE heat pumps in the back and bottom cushions. The TE system makes conditioned flowing air through channels to the occupant for providing on-demand cooling or heating. As shown in the first panel in **Figure 6**, the seat has the heat pump consisting of a TE module (green box) and a fan (orange).

- Thermal management of tiny laser diodes is used in fiber optic telecom, datacom backhaul networks. TEC can also be used for contact cooling of semiconductor lasers, infrared detectors, CCD- matrix, and miniconditioners for photomultipliers.

- Localized cooling at hot spots of chips was created. For example, the Intel group is the first to demonstrate both concepts of applying the TE material only to a chip's hottest spots (**Figure 6**) [33, 37]. On the substrate, the researchers grew a 100-μm-thick layered structure, called a superlattice, containing bismuth, tellurium, antimony, and selenium. The structure can pump 1300W/cm^2 heat from the back side of the chip to the heat spreader. The superlattice induced an approximately 6°C temperature drop at the hot spot even before the device was powered up, because it conducts heat better than the grease that bonds the rest of the heat spreader to the chip. Yet, when a 3 A-current went through the thermoelectric cooler, the total temperature change was only of 15°C. Managing heat in electronics is a common issue, and TE

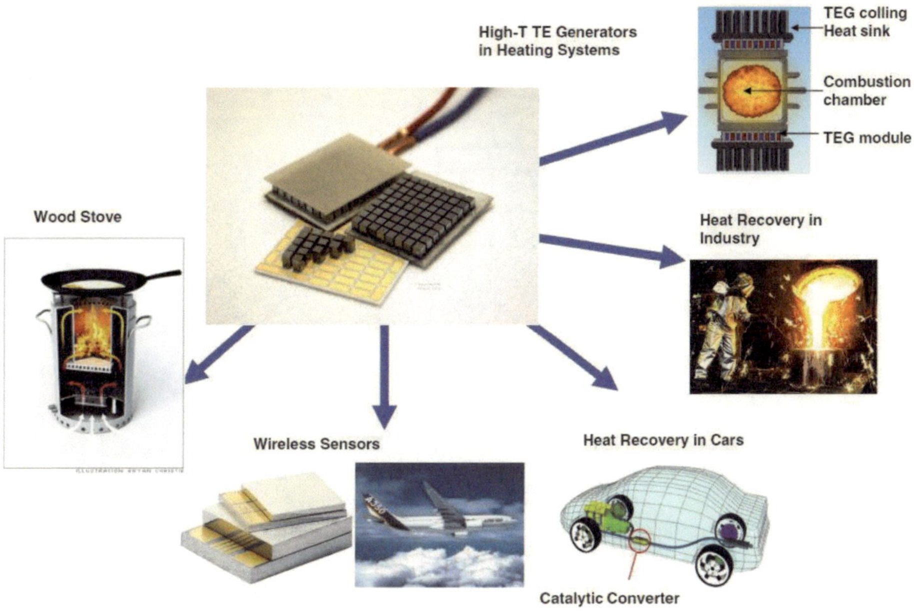

Figure 7.
Overview of potential thermoelectric generator (TEG) applications [33, 34].

coolers can improve electronic systems in thermal performance, cost, noise, weight, size or efficiency.

Figure 7 shows an overview of the present and potential applications of thermoelectric generators (TEGs) [34]. They include (1) heating systems and water boilers with TEG units which generate the electricity for the control units and pumping systems, (2) the long term perspective of waste heat recovery for medium-scale industrial facilities, (3) waste heat recovery in automobiles and other combustion-engine-powered vehicles for enhanced efficiency and electric current supply of the electronic system, (4) miniaturized autarkic sensor systems powered by an integrated TEG with a wireless data transmitter, (5) ventilated wood stove powered by a thermoelectric generator with enhanced oxygen supply, improves burning process.

2. Nanostructured Bi_2Te_3-based thermoelectric thin films grown using pulsed laser deposition

2.1 PLD growths of nanostructured Bi_2Te_3-based thin films

PLD is one of the most convenient thin film growth techniques that uses a high intensity pulsed laser beam as an external energy source to ablate a target, form a plume, and deposit thin films onto a substrate. In practice, a large number of variables affect the properties of the film, such as substrate temperature (T_s), background gas pressure (P) and laser fluence. **Figure 8** shows a PLD system for preparing thermoelectric thin films [38, 39]. The substrate was heated and maintained at desired T_s using a thermocouple and a proportional-integral-derivative temperature controller. The thermocouple was buried inside a substrate holder which was heated by a tungsten lamp or electrical resistance heating. The pressure of ambient gas (He, Ar) could be fine-tuned by the needle valve. Laser source can

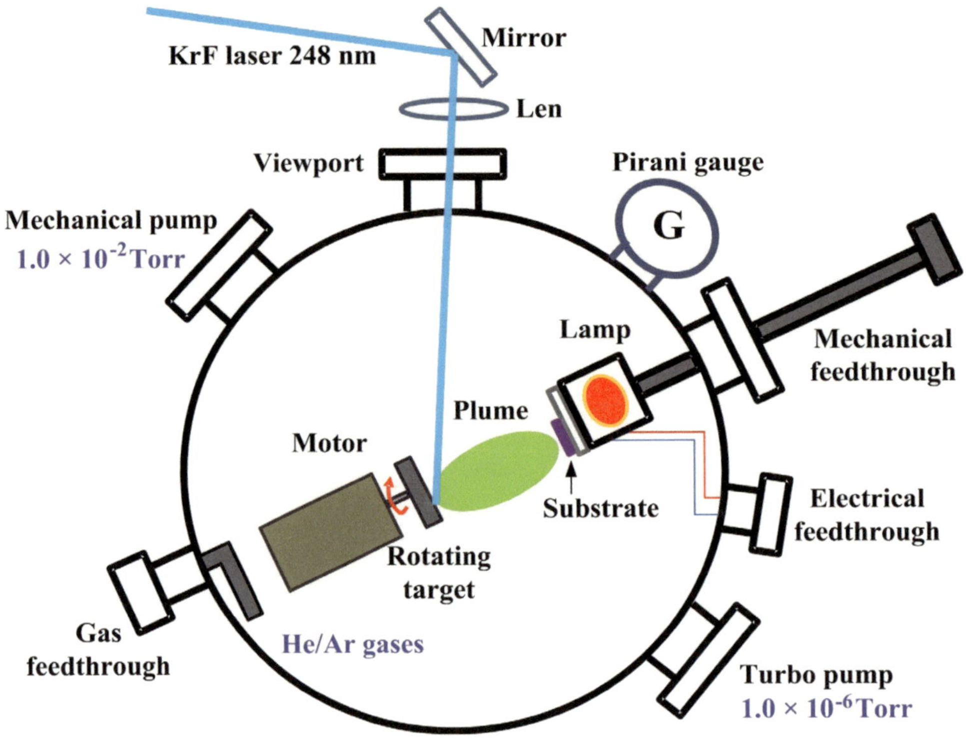

Figure 8.
A schematic illustration of a PLD system.

be KrF excimer laser beam (λ = 248 nm) and Q-switched Nd:YAG laser (λ = 355 nm) with properly selected laser fluence (e.g., 3.8, 6.2, or 8.3 J/cm^2) pulsed duration of 5–20 ns, repetition rate of 5–10 Hz [38, 40–42]. The laser beam was guided by several UV mirrors and focused on a stoichiometric polycrystalline target (e.g., Bi_2Se_3, Bi_2Te_3, $Bi_{0.5}Sb_{1.5}Te_3$, etc.) inside the vacuum chamber by the UV lens. The deposition chamber was evacuated to a base pressure of $\sim 10^{-6}$ Torr, and high-purity ambient gas (He or Ar) was then introduced until obtaining a target pressure (e.g., usually $10^{-5} - 3\times10^{-1}$ Torr).

The enhancement of the PF of Bi_2Te_3-based thin films is challenging due to the coupling among TE material properties [3], and the difficulty in growing stoichiometric films [38]. Indeed, stoichiometry is a key factor for obtaining better TE properties [5, 38, 43–45]. Yet, both tendency for re-evaporation of volatile elements (i.e., Te, Se) at elevated T_s [45–48] and the low sticking coefficient Te (< 0.6 for Bi_2Te_3) at T_s beyond 300°C [49, 50] constrain to grow stoichiometric Bi_2Te_3-based films (**Figure 9a** and **b**).

2.2 Super-assembling of Bi_2Te_3 hierarchical nanostructured thin films

C.-H. Chen et al. [41] reported the PLD growths of super-assembling of Bi_2Te_3 hierarchical nanostructured thin films on the SiO_2/Si substrates and their thermoelectric properties. Interesting Bi_2Te_3 super-assemblies were successfully grown using PLD with controlling the substrate temperatures from 350–600°C and at a fixed Ar ambient pressure of approximately 10^{-3} Torr. SEM images in **Figure 10** clearly shows the morphological characteristics of the superassembling Bi_2Te_3 nanostructured thin films [41]. At lower deposition temperatures (< 450°C), the films are mainly composed of vertically aligned nanoscaled flakes, but flakes are

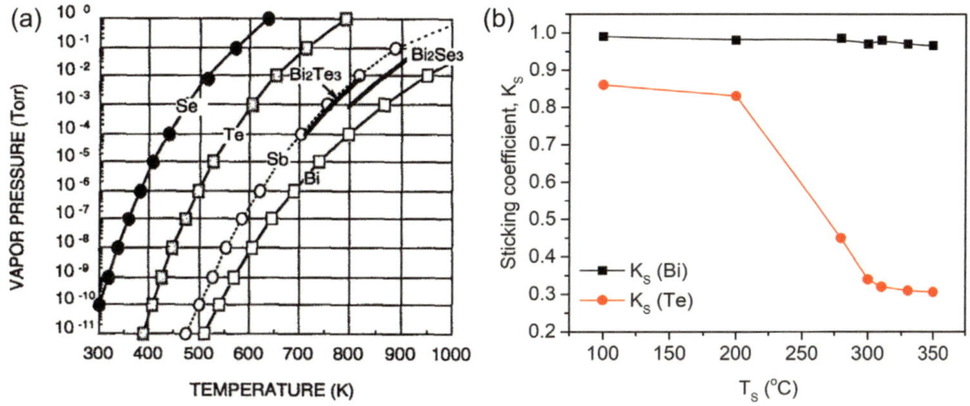

Figure 9.
(a) Vapor pressures of Bi, Sb, Te, Se, Bi_2Se_3, and Bi_2Te_3 as a function of temperature [46]. (b) The variation of sticking coefficient K_s (Bi, Te) as a function of substrate temperature T_s at fixed flux ratio F_R = 4.5 [49].

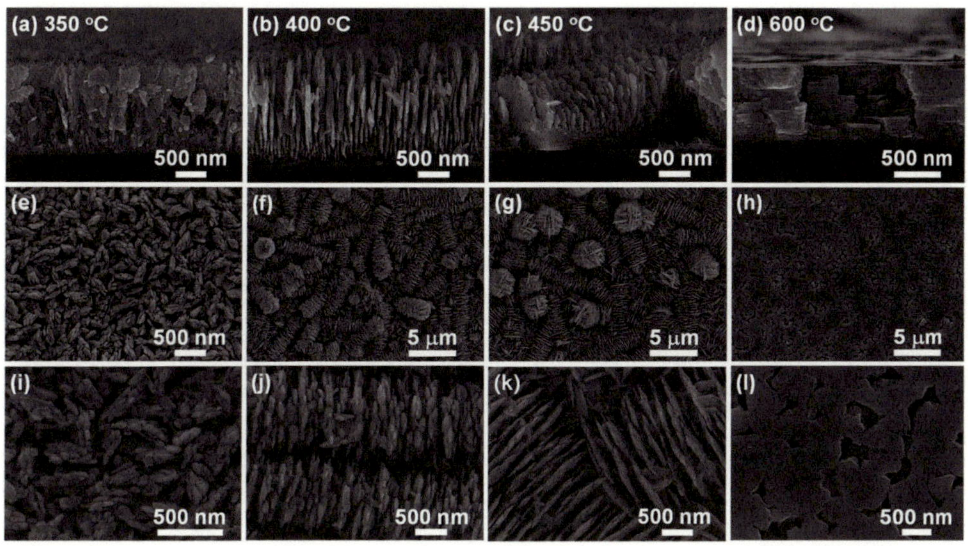

Figure 10.
(a)–(d) The cross-sectional and (e)–(l) the corresponding top-view SEM images of the Bi_2Te_3 superassemblies deposited at 350°C, 400°C, 450°C, and 600°C, respectively [41].

horizontally stacked for 600°C-film (**Figure 10a–d**). Moreover, the bottom of each of the deposited super-assemblies has a relatively continuous and dense layer, and this layer thickness increases with increasing substrate temperature from 350–450° C (**Figure 10a–d**). The top-view SEM images confirm for the high uniformity and presents the unique super-assembling features of the repetitively and regularly assembled nano-flakes (**Figure 10e–h**). These four films are uniformly composed of spindle-like (**Figure 10e**), worm-like (**Figure 10f** and **g**) and island-like (**Figure 10h**) hierarchical nanostructures. Magnified top-view SEM images (**Figure 10i–k**) further show that the nanoflakes are composed of oriented and regular assemblies of numerous rice-like and elongated primitive nanoparticles [41]. At a higher substrate temperature, thin- and large-size nanoflakes are formed from packing of dense rice-like nanoparticles, driving by the relatively sufficient thermal energy for diffusion. In addition, the out-of- plane superassembly structure (600°C) has a limited column width, which is not always consistent along the out-of-plane direction (**Figure 10d**). Also, the parallel nano-flakes (at 600°C) are evidently formed by flake stacking along c-axis orientation or epitaxial-like growth. The special three-dimensional mesh-like structure of 600°C-film would also be an effective design for scattering phonons, and it's extremely smooth top surface is certainly beneficial for subsequent analyses and applications [41].

Figure 11(a) shows the crystal structure of Bi_2Te_3, which is usually described by a hexagonal cell that consists of 15 layers of atoms stacking along the c-axis with a sequence [5], namely $\cdots Te^{(1)}–Bi–Te^{(2)}–Bi–Te^{(1)} \cdots Te^{(1)}–Bi–Te^{(2)}–Bi–Te^{(1)} \cdots Te^{(1)}–Bi–Te^{(2)}–Bi–Te^{(1)} \cdots$. The superscripts refer to two different types of bonding for Te atoms. The 5-atomic-layer thick lamellae of–($Te^{(1)}–Bi–Te^{(2)}–Bi–Te^{(1)}$)– is called quintuple layers, QLs. The $Te^{(1)}$ ... $Te^{(1)}$ refers Van der Waals force between Te atoms, whereas the $Te^{(1)}$–Bi and Bi–$Te^{(2)}$ are ionic-covalent bonds. This weak binding between the $Te^{(1)}$ ... $Te^{(1)}$ accounts for the anisotropic thermal and electrical transport properties of Bi_2Te_3. For example, the thermal conductivity along the *c*-axis direction ($\sim$0.7 $Wm^{-1}K^{-1}$) is approximately a haft of the value along the plane perpendicular to the *c*-axis ($\sim$1.5 $Wm^{-1}K^{-1}$) [5, 6, 13]. The weak binding of $Te^{(1)}$ ... $Te^{(1)}$ also make the ease of cleavage along the plane perpendicular to the *c*-axis.

Figure 11(b) shows XRD patterns of the Bi_2Te_3 super-assemblies deposited at various substrate temperatures from 350–600°C. Clearly, all the films exhibited rhombohedral Bi_2Te_3 (JCPDS no. 89–4302) without traceable impurities or oxides. When substrate temperature increases, the (00 l) preferential orientation gradually becomes stronger, the 600°C- film is highly (00 l)-preferred orientation, which is consistent with the SEM observation (**Figure 10d**). The gradually enhanced (00 l) peaks from 350–450°C mainly originate from the increased thickness of the bottom layer (**Figure 10a–d**), which has similar lamellar morphology with (00 l)-preferential orientation of 600°C-film [41].

Figure 12 presents the proposed growth model of the super-assemling nano-structured Bi_2Te_3 films prepared at various T_S. The growth mechanisms are layer-then-fake for T_S = 350–450°C and layer-by-layer for higher T_S of 600°C. We can only find a monotone morphology and single preferential orientation of (00 l) for 600°C-film, which lead to a fully lamellar morphology with the (00 l) preferential

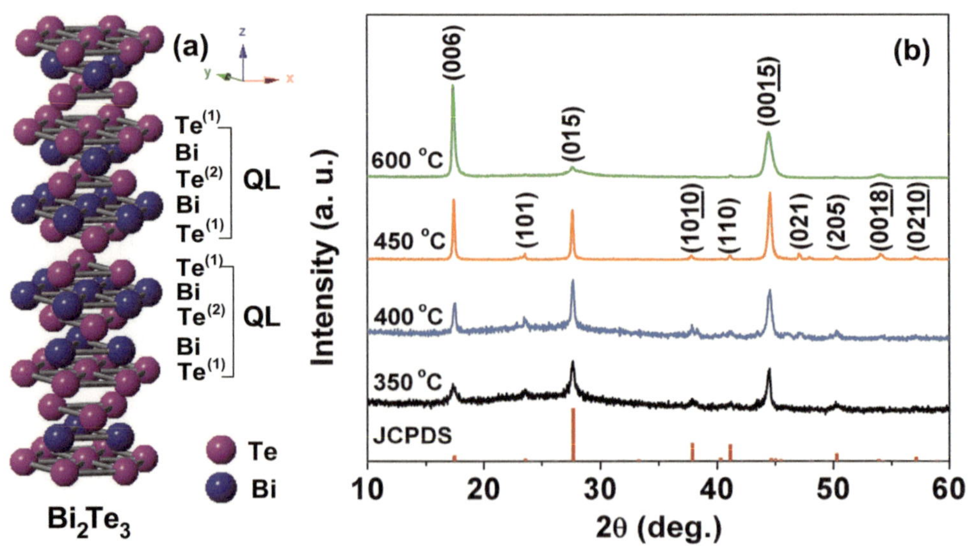

Figure 11.
(a) Crystal structures of Bi_2Te_3. (b) X-ray diffraction (XRD) patterns of of the Bi_2Te_3 super-assemblies deposited at various deposition temperatures from 350–600°C and at an Ar ambient pressure $\sim 10^{-3}$ Torr [41].

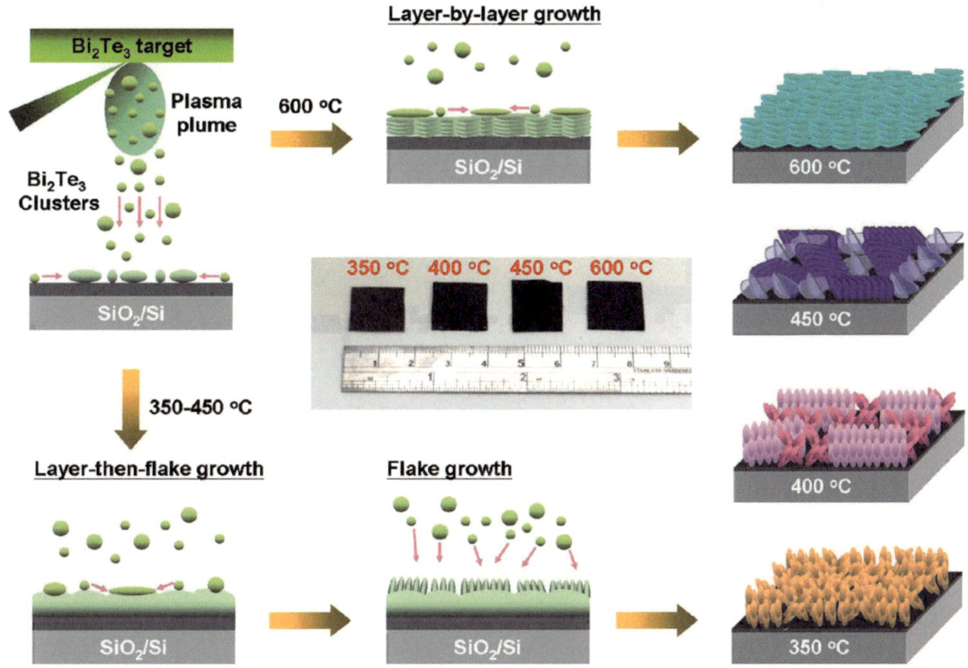

Figure 12.
Schematic illustration of the layer-then-flake and layer-by-layer growth models and the resulting Bi_2Te_3 in-plane (350–450°C) and out- of-plane (600°C) super-assemblies. Inset is the optical image of the prepared super-assembled films with a size of 1.5 × 1.5 cm^2 [41].

orientation. Meanwhile, drastic changes in morphologies from layer to flake and orientations from (00 l) to (015) are observed at lower temperatures (350–450°C). The (00 l)-preferred orientation should be attributed to the thin bottom layer of the films prepared at 350–450°C. The thickness of this layer increases with increasing T_S. Since the bottom layer at 350°C is extremely thin, the required *Ts* for obtaining layer growth should be just below 350°C. The drastic change in the morphology and orientation at T_S of 350–450°C, namely, the layer- then-flake growth can be induced by a temperature gradient along the growth direction that the temperature at top surface of the as-deposited film should be slightly lower than at the substrate [41].

Table 1 summarizes the detailed properties of the super-assembling nanostructured Bi_2Te_3 thin-films. Due to such the voided structures, the films exhibited low electrical conductivity from 49 $S.cm^{-1}$ for worm-like superassembly (450°C) to 160 $S.cm^{-1}$ for 3D-layered super-assembly (600°C). Seebeck coefficient of the films was in range of 113–138 μV/K. As a result, the power factor (PF) is relatively low in range of 0.93 to 3.0 $\mu W/cmK^2$, primarily due to the low electrical conductivity of the films with voided morphologies.

In PLD, tightly controlling substrate temperatures (*Ts*) and ambient pressures (*P*) enables the morphologies and compositions of films to be manipulated extensively, which offers a new method for enhancing the TE properties of films [38, 43, 51, 58, 59]. For example, self-assembled Bi_2Te_3 films featuring well-aligned zero- to three-dimensional nanoblocks have been fabricated (**Figure 13a–d**), but the room-temperature PFs of these films remain low ($\leq 1.9\ \mu Wcm^{-1}\ K^{-2}$) [51]. By contrast, A. Li Bassi et al. [43] obtained several microstructured Bi_2Te_3 films (**Figure 13e–h**) with high PFs for morphologies: layered-smooth (50.6 $\mu Wcm^{-1}\ K^{-2}$, **Figure 13e, e1**), and compact-smooth (21.2 $\mu Wcm^{-1}\ K^{-2}$, **Figure 13f,f1**) at room-temperature;

Material	Morphology	Method	Deposition conditions		n (10^{19} cm^{-3})	μ (cm^2/Vs)	σ (S/cm)	α (μV/K)	PF (μW/cmK^2)	Ref.
Bi_2Te_3	Spindle-like super-assembly	PLD	350°C	0.13 Pa	4.0	12.4	79	−113	1.01	[41]
Bi_2Te_3	Worm-like super-assembly	PLD	400°C	0.13 Pa	1.9	25.9	73	−119	1.03	[41]
Bi_2Te_3	worm-like super-assembly	PLD	450°C	0.13 Pa	1.2	29.4	49	−138	0.93	[41]
Bi_2Te_3	3D-layered super-assembly	PLD	600°C	0.13 Pa	5.1	20.3	160	−137	3.0	[41]
Bi_2Te_3	Nanoparticles	PLD	300°C	20 Pa	9.7	14.8	230	−91	1.90	[51]
Bi_2Te_3	Nanoparticles	PLD	300°C	1.0 Pa	105	8.3	1390	−60	5.0	[52]
Bi_2Te_3	Layered Structure	Sputtering	350°C	1.0 Pa	95	12.1	1840	−70	8.8	[52]
	Nanorods		250°C	0.9 Pa	9.1	2.0	29	−81	0.19	[51]
Bi_2Te_3	Columnar Structure	Sputtering	350°C	1.0 Pa	246	7.5	2990	−46	6.4	[52]
Bi_2Se_3	Layered HPs	PLD	300°C	40 Pa	7.4	81.4	963.8	−75.8	5.5	[38]
Bi_2Te_3	Compact film	PLD	300°C	80 Pa	5	102	814.3	−172.8	24.3	[40]
Bi_2Te_3	Layered-smooth film	PLD	250°C	10 Pa	10.1	90.6	1464	−186	50.6	[43]
Bi_3Se_2Te	Nanocrystalline film	PLD	250°C	40 Pa	35.5	34.4	1747.5	−68.8	8.3	[53]
Bi_2Se_3	Bulk	Melting and hot-pressing			—	—	251.9	−175	7.7	[54]
$Bi_2Se_{0.3}Te_{2.7}$	Bulk	Ball milling-hot pressing			—	—	892	−190	32.2	[55]
$Bi_2Se_{1.5}Te_{1.5}$	Bulk	Zone melting			1.2	230	441.6	−193	16.5	[56]
$Bi_2Se_{1.8}Te_{1.2}$	Nano-platelet bulk	Polyol method			—	—	199.6	−80.9	1.3	[57]
Bi_2Se_2Te	Bulk	Ball milling- hot pressing			—	—	1613	−60	5.8	[55]

Table 1.
Material, type, method, processing conditions, carrier concentration (n), mobility (μ), electrical conductivity (σ), Seebeck coefficient (α), power factor ($PF = \alpha^2\sigma$) of the Bi_2Te_3 films deposited by PLD and RF sputtering, as compared to properties of Bi_2Se_3, Bi_2Se_3, $Bi_2Se_xTe_{1-x}$ bulk and films reported in the literature. All the selected values were recorded at room temperature.

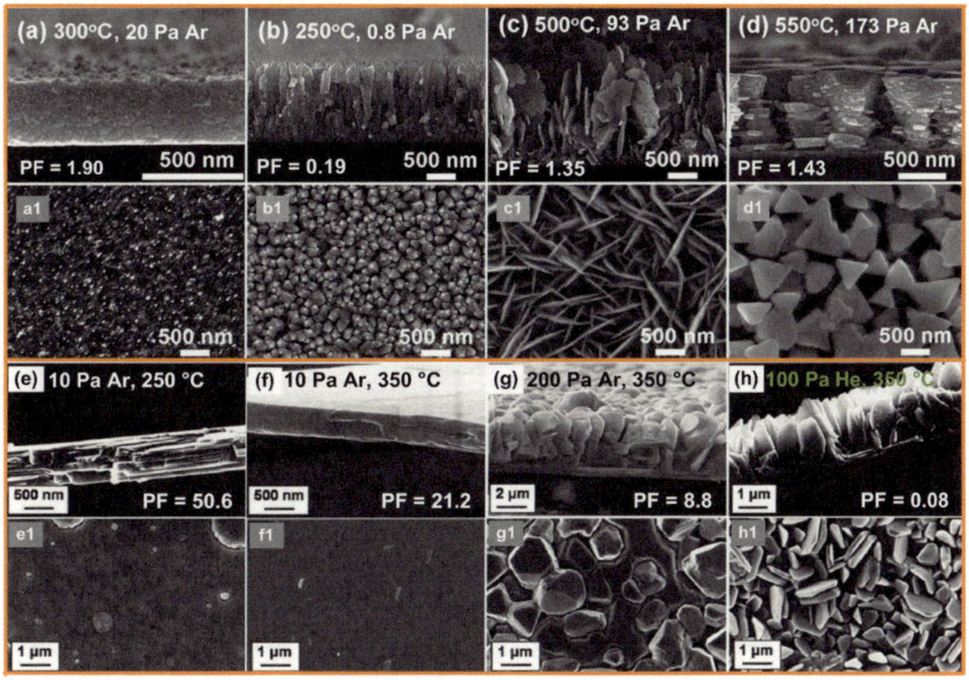

Figure 13.
The morphology and power factor (unit $\mu Wcm^{-1} K^{-2}$) of nano/micro-structured Bi_2Te_3 thin-films grown by PLD at various substrate temperatures and ambient pressures, reported by (i) Chang and Chen [51] and (ii) Li Bassi et al. [43].

whereas the PFs remained low values of 8.8 $\mu Wcm^{-1} K^{-2}$ for 3D crystallite shapes (**Figure 13g,g1**) and 0.08 $\mu Wcm^{-1} K^{-2}$ for 3D-voided platelets (**Figure 13h,h1**).

Table 1 summaries the morphology and properties of Bi_2Te_3-based thin-films deposited by PLD, sputtering, in comparison with the properties of TE bulks. Usually, TE nanomaterials possess low σ values due to the separating or voided structure-morphology, but bulk and thin films have superior σ. For example, the compact-polycrystalline Bi_3Se_2Te achieved σ = 1747.5 S/cm [53] or even higher for some other films [52]. Unfortunately, the κ of films are missed in many published works to calculate ZT of the films. Thermal conductivity (κ) of nanocrystalline and nanostructured thermoelectrics is expected to achieve low values thanks to the extensive phonons scattering at interfaces, surfaces and grain boundaries. Indeed, reduced κ values have been noted for the monocrystalline $Bi_2Se_{0.3}Te_{2.7}$ films (κ = 0.8 W/m K for an average grain size of 60 nm) [60], and for Bi-Sb-Te films [61, 62].

2.3 Thermal conductivity κ of Bi_2Se_3 and Bi_2Te_3 and Bi-Te-Se compounds

A transient 3ω technique is usually employed in measuring thermal conductivity of thermoelectric films. The detail of this technique can be found in refs. [62–64]. **Table 2** summarizes thermal transport properties (at room–temperature) of nano-crystalline–nanostructured Bi_2Te_3-based thin films and bulk materials in the literature. Generally, the thermal conductivity κ value for polycrystalline films is expected to be smaller than that of bulk alloys because of the extensive phonons scattering at interfaces, surfaces and grain boundaries [5, 60, 66]. Moreover, the κ of nanocrystalline Bi_2Te_3-based films will further decrease when the grain size of

Sample, fabrication method	Avg. grain size	κ (W/m K)	σ (S/cm)	α (μV/K)	PF = $\sigma\alpha^2$ (μW/cmK^2)	ZT (300 K)	Ref.
$Bi_2Te_{2.7}Se_{0.3}$ nanocrystalline thin film, flash evaporation	60 nm	0.8 (cross-plane)	540	−186.1 (in-plane)	18.7 (in-plane)	0.7	[60]
Sintered bulk $Bi_2Te_{3-x}Se_x$ material, hot-pressing	30 μm	1.6	930	−177.5	29.3	0.6	
Nanocrystalline bismuth-telluride-based ($Bi_2Te_{3-x}Se_x$) thin film	10 nm	0.61	550	−84.0	3.9	0.19	[65]
	27 nm	0.68	540	−138.1	10.3	0.46	
	60 nm	0.80	540	−186.1	18.7	0.70	
Nanocrystalline Bi-Sb-Te thin film, sputtering	26 nm	0.46	3.3	—	—	—	[62]
	45 nm	0.65	6.7	—	—	—	
	84 nm	0.81	33.3	—	—	—	
Nanocrystalline BiSbTe (8:30:62) thin film, flash evaporation	150 nm	0.6	—	—	—	—	[66]
Single crystal BiSbTe bulk alloys	—	0.75	—	—	—	—	[5]
Bi_2Te_3/Sb_2Te_3 superlattices (period~5 nm)	—	0.4	—	—	—	—	[67]
$Bi_2Te_{3+0.63}$ bulk	—	2.2	1000	−240	58	0.87	[5]
$Bi_2(Te_{0.95}Se_{0.05})_3$ bulk	—	1.59	901	−223	45	0.85	[5]
$Bi_2Te_3/Bi_2(Te_{0.88}Se_{0.12})_3$ superlattice film, MBE	80 nm	1.25	639	−204	27	0.60	[68]
Bi_2Te_3 film, PLD	—	1.1	—	—	—	—	[69]
Bi_2Te_3/Sb_2Te_3 superlattices film (layered thickness ~ 6 nm), PLD.	—	0.11	—	—	—	—	
Bi_2Te_3 films, laser ablation	—	0.2–0.3	—	—	—	—	[70]
$Bi_xSb_{2-x}Te_3$ nanolayer film, PLD	190 nm	1.16	2700	95	25	0.65	[71]
$Bi_xSb_{2-x}Te_3$ nanodisc film, PLD	100 nm	1.00	1100	132	20	0.60	
$Bi_xSb_{2-x}Te_3$ nanocolumn film, PLD	70 nm	0.93	280	207	12	0.39	

Table 2.
Room–temperature thermal transport properties of nanocrystalline–nanostructured Bi_2Te_3-based thin films and bulk materials in the literature, included: sample and fabrication method, average grain size, thermal conductivity κ, electrical conductivity σ, Seebeck coefficient α, power factor PF (= $\alpha^2\sigma$), and ZT (at 300 K).

decreases (κ ≤ 0.81 W/mK, **Figure 14A**) [62, 65]. For Bi_2Te_3/Sb_2Te_3 superlattice films, the coherent backscattering of phonon waves at the superlattice interfaces is outlined for the reduction of lattice thermal conductivity, resulting in the low κ ≤ 0.4 W/mK [67, 69].

For PLD Bi_2Te_3-based films, Yamasaki et al. [69] measured thermal conductivity with an ac calorimetric method in the direction across the film, obtaining k~1.1 W/m K for a Bi_2Te_3 film deposited by PLD in vacuum (**Table 2**). In addition, Walachova et al. [70] used direct ZT measurement with the Harman method to estimate the κ value, and it is about 0.2–0.3 W/mK for the Bi_2Te_3 films. Recently, Chang et al. [71] reported the κ values of 0.93–1.16 W/mK for $Bi_xSb_{2-x}Te_3$ films with the granular-layered morphologies (**Figure 14B**).

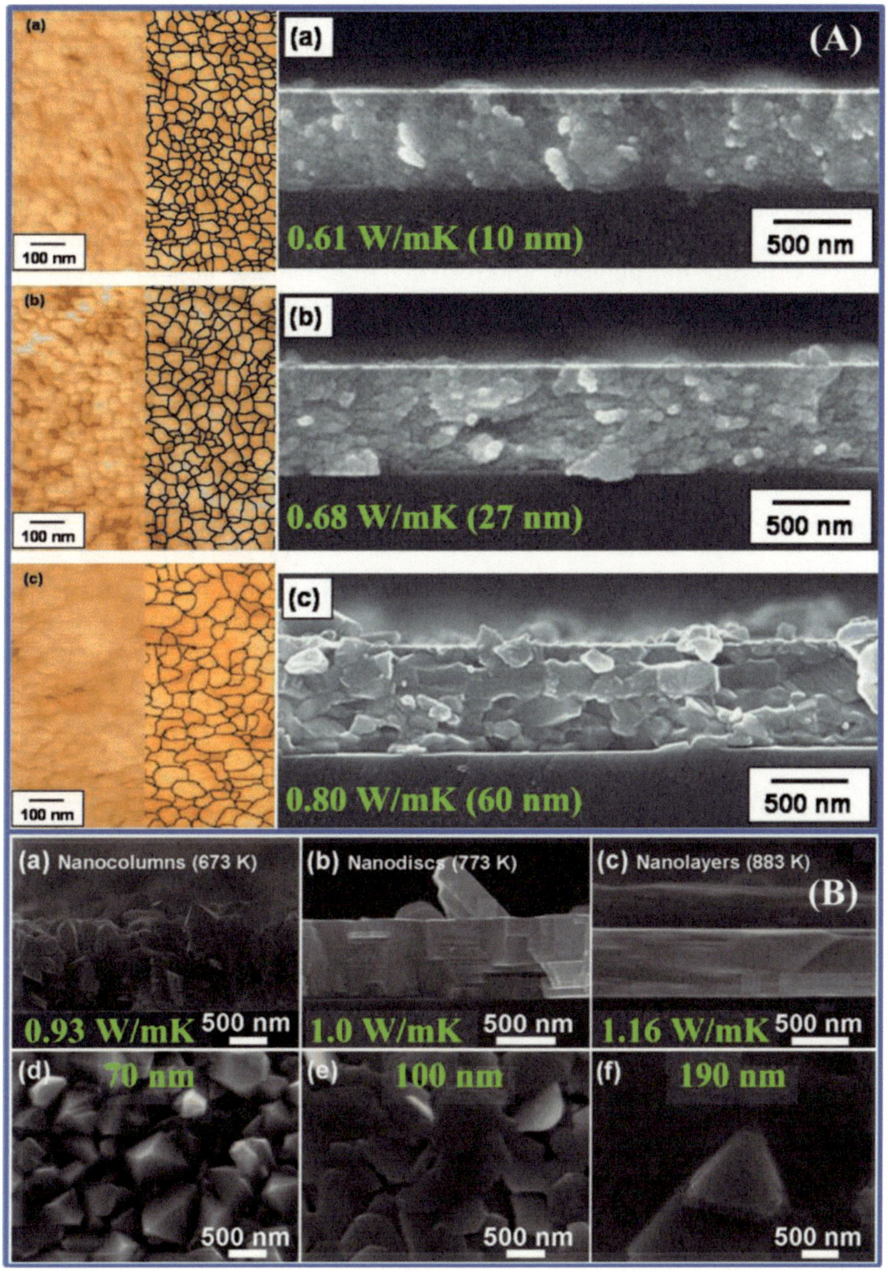

Figure 14.
The morphology and thermal conductivity of Bi_2Te_3-based films with different grain sizes: (A) nanocrystalline $Bi_2Te_{3-x}Se_x$ films [65], (B) the $Bi_xSb_{2-x}Te_3$ films [71].

3. Conclusions

In this book chapter, we present an overview of thermoelectric materials and applications, challenging of enhancing TE properties, and the nanostructuring approach in development TE materials. Various interesting nanostructured Bi_2Te_3-based thin films have been grown successfully by PLD with properly controlled substrate temperatures ambient gas pressures. For example, super-assembling of

Bi_2Te_3 hierarchical nanostructures were grown at T_S from 350 to 600°C, and the films possessed relative high Seebeck coefficient of 113–138 μV/K, but exhibited low electrical conductivities of 49–160 $S.cm^{-1}$, and thus they had relatively low PF in range of 0.93 to 3.0 $\mu W/cmK^2$. Our intensive literature review on Bi_2Te_3-based TE materials can make general conclusion that TE nanomaterials possess low σ values when their structure-morphology are separating or voided, meanwhile, bulk and compact-smooth thin films can achieve high σ values. The PF values of Bi_2Te_3-based thermoelectrics varied in a wide range, i.e. below 5.0 $\mu W/cmK^2$ for voided structure-morphology, and reaching intermediate-high PF values of 5.0–50.6 $\mu W/cmK^2$ for compact-smooth or compact-layered structures. An advantage of nanocrystalline and nanostructuring thermoelectrics is the reduced thermal conductivity (possibly below 1 $W.m^{-1}\ K^{-1}$). This book chapter provides fundamental understanding the relationship amongst processing condition in PLD growths, structure-morphology, and TE properties of Bi_2Te_3-based thin films.

Acknowledgements

This research is funded by Vietnam National Foundation for Science and Technology Development (NAFOSTED) under grant number 103.02-2019.374, and the Ministry of Science and Technology, Taiwan under Contract Nos. MOST 109-2221-E-214-016 and MOST 110-2221-E-214-013.

Author details

Le Thi Cam Tuyen[1], Phuoc Huu Le[2]* and Sheng-Rui Jian[3,4]*

1 Department of Physics, Faculty of Basic Sciences, Nam Can Tho University, Can Tho City, Vietnam

2 Department of Physics and Biophysics, Faculty of Basic Sciences, Can Tho University of Medicine and Pharmacy, Can Tho City, Vietnam

3 Department of Materials Science and Engineering, I-Shou University, Kaohsiung, Taiwan

4 Department of Fragrance and Cosmetic Science, College of Pharmacy, Kaohsiung Medical University, Kaohsiung, Taiwan

*Address all correspondence to: lhuuphuoc@ctump.edu.vn and srjian@gmail.com

IntechOpen

© 2021 The Author(s). Licensee IntechOpen. This chapter is distributed under the terms of the Creative Commons Attribution License (http://creativecommons.org/licenses/by/3.0), which permits unrestricted use, distribution, and reproduction in any medium, provided the original work is properly cited. CC BY

References

[1] Bell LE. Cooling, heating , generating power, and recovering waste heat with thermoelectric systems. Science. 2008; **321**: 1457–1461. DOI: 10.1126/science.1158899

[2] Chen G., Shakouri A. Heat Transfer in Nanostructures for Solid-State Energy Conversion. J. Heat Transfer. 2002; **124**: 242. DOI: 10.1115/1.1448331.

[3] Snyder GJ, Toberer ES. Complex thermoelectric materials. Nat. Mater. 2008; 7:105–114. DOI: 10.1038/nmat2090

[4] Li J-F, Liu W-S, Zhao L-D, Zhou M. High-performance nanostructured thermoelectric materials. NPG Asia Mater. 2010; **2**: 152. DOI: 10.1038/asiamat.2010.138

[5] Rowe DM. Thermoelectrics handbook: macro to nano. D.M. Rowe Ed. Boca Raton: CRC Press; 2006. 1014 p. DOI: 10.1201/9781420038903

[6] Nolas GS, Sharp J, Goldsmid HJ, Thermoelectrics: Basic Principles and New Materials Developments. New York: Springer; 2001.

[7] Lan Y, Minnich AJ, Chen G, Ren Z. Enhancement of thermoelectric figure-of-merit by a bulk nanostructuring approach. Adv. Funct. Mater. 2010; **20**: 357–376. DOI: 10.1002/adfm.200901512

[8] http://www.electronics-cooling.com/2005/11/advances-in-high-performance-cooling- for-electronics/

[9] Nolas GS, Morelli DT, Tritt TM. KUTTERUDITES: A Phonon-Glass-Electron Crystal Approach to Advanced Thermoelectric Energy Conversion Applications. Annu. Rev. Mater. Sci. 1999; **29**: 89. DOI: 10.1146/annurev.matsci.29.1.89.

[10] Alam H and Ramakrishna S. A review on the enhancement of figure of merit from bulk to nano-thermoelectric materials. Nano Energy. 2013; **2**: 190–212. DOI: 10.1016/j.nanoen.2012.10.005

[11] Pichanusakorn P, Bandaru P. Nanostructured thermoelectrics. Mater. Sci. Eng. R. 2010; **67**:19–63. DOI: 10.1016/j.mser.2009.10.001

[12] Cutler M, Leavy JF, Fitzpatrick RL. Electronic transport in semimetallic cerium sulfid. Phys. Rev. 1964; **133:** A1143.

[13] Tritt TM. Thermoelectric Phenomena, Materials, and Applications. Annu. Rev. Mater. Res. 2011; **41:** 433. DOI: 10.1146/annurev-matsci-062910-100453.

[14] Martín-González M, Caballero-Calero O, Díaz-Chao P. Nanoengineering thermoelectrics for 21st century: Energy harvesting and other trends in the field. *Renew. Sustain. Energy Rev.* 2013; **24**: 288. DOI: 10.1016/j.rser.2013.03.008.

[15] Eilertsen J, Li J, Rouvimov S, Subramanian MA. Thermoelectric properties of indium-filled $In_xRh_4Sb_{12}$ skutterudites. J. Alloys Compd. 2011; **509**: 6289. DOI: 10.1016/j.jallcom.2011.03.057.

[16] Dresselhaus MS, Chen G, Tang MY, Yang RG, Lee H, Wang DZ, Ren ZF, Fleurial J-P, Gogna P. New Directions for Low-Dimensional Thermoelectric Materials. Adv. Mater. 2007; **19**: 1043. DOI: 10.1002/adma.200600527.

[17] Hicks LD, Dresselhaus MS. Effect of quantum-well structures on the thermoelectric figure of merit. Phys. Rev. B 1993; **47:** 12727. DOI: 10.1103/PhysRevB.47.12727

[18] Hicks LD, Dresselhaus MS. Thermoelectric figure of merit of a one-dimensional conductor. Phys. Rev. B

1993; **47:** 16631. DOI: 10.1103/PhysRevB.47.16631

[19] Heremans JP, Jovovic V, Toberer ES, Saramat A, Kurosaki K, Charoenphakdee A, Yamanaka S, Snyder GJ. Enhancement of thermoelectric efficiency in PbTe by distortion of the electronic density of states. Science. 2008; **321**: 554–557. DOI: 10.1126/science.1159725

[20] Vineis CJ, Shakouri A, Majumdar A, Kanatzidis MG. Nanostructured thermoelectrics: big efficiency gains from small features. Adv. Mater. 2010; **22:** 3970. DOI: 10.1002/adma.201000839.

[21] Androulakis J, Lin C-H, Kong H-J, Uher C, Wu C-I, Hogan T, Cook BA, Caillat T, Paraskevopoulos KM, Kanatzidis MG. Spinodal decomposition and nucleation and growth as a means to bulk nanostructured thermoelectrics: enhanced performance in $Pb_{1-x}Sn_xTe$-PbS. J. Am. Chem. Soc. 2007; **129**: 9780. DOI: 10.1021/ja071875h.

[22] Rhyee J-S, Lee KH, Lee SM, Cho E, Kim SI, Lee E, Kwon YS, Shim JH, Kotliar G. Peierls distortion as a route to high thermoelectric performance in In(4)Se(3-delta) crystals. Nature 2009; **459**: 965. DOI: 10.1038/nature08088.

[23] Saramat A, Svensson G, Palmqvist AEC, Stiewe C, Mueller E, Platzek D, Williams SGK, Rowe DM, Bryan JD, Stucky GD. Large thermoelectric figure of merit at high temperature in Czochralski-grown clathrate $Ba_8Ga_{16}Ge_{30}$. J. Appl. Phys. 2006; **99**: 023708. DOI: 10.1063/1.2163979.

[24] Wang XW, Lee H, Lan YC, Zhu GH, Joshi G, Wang DZ, Yang J, Muto AJ, Tang MY, Klatsky J, Song S, Dresselhaus MS, Chen G, Ren ZF. Enhanced thermoelectric figure of merit in nanostructured n-type silicon germanium bulk alloy. Appl. Phys. Lett. 2008; **93**: 193121. DOI: 10.1063/1.3027060.

[25] Venkatasubramanian R, Siivola E, Colpitts T, O'Quinn B. Thin-Film thermoelectric devices with high room-temperature figures of merit. Nature. 2001; **413**: 597–602. DOI: 10.1038/35098012

[26] Hsu KF, Loo S, Guo F, Chen W, Dyck JS, Uher C, Hogan T, Polychroniadis EK, Kanatzidis MG. Cubic $AgPb_mSbTe_{2+m}$: bulk thermoelectric materials with high figure of merit. Science. 2004; **303**: 818. DOI: 10.1126/science.1092963.

[27] Chung DY, Hogan T, Brazis P, Rocci-Lane M, Kannewurf C, Bastea M, Uher C, Kanatzidis MG. $CsBi_4Te_6$: A High-Performance Thermoelectric Material for Low-Temperature Applications. Science 2000; **287**: 1024. DOI: 10.1126/science.287.5455.1024.

[28] Harman TC, Taylor PJ, Walsh MP, LaForge BE. Quantum dot superlattice thermoelectric materials and devices. Science. 2002; **297**: 2229. DOI: 10.1126/science.1072886.

[29] Boukai AI, Bunimovich Y, Tahir-Kheli J, Yu J-K, Goddard WA, Heath JR. Silicon nanowires as efficient thermoelectric materials. Nature. 2008; **451**: 168. DOI: 10.1038/nature06458.

[30] Hochbaum AI, Chen R, Delgado RD, Liang W, Garnett EC, Najarian M, Majumdar A, Yang P. Enhanced thermoelectric performance of rough silicon nanowires. Nature 2008; **451**: 163. DOI: 10.1038/nature06381.

[31] Poudel B, Hao Q, Ma Y, Lan Y, Minnich A, Yu B, Yan X, Wang D, Muto A, Vashaee D, Chen X, Liu J, Dresselhaus MS, Chen G, Ren Z. High-thermoelectric performance of nanostructured bismuth antimony telluride bulk alloys. Science. 2008: **320:** 634. DOI: 10.1126/science.1156446.

[32] Poudeu PFP, Angelo JD', Downey AD, Short JL, Hogan TP, Kanatzidis MG. High thermoelectric figure of merit and nanostructuring in bulk p-type $Na_{1-x}Pb_mSb_yTe_{m+2}$. Angew. Chem. Int. Ed. Engl. 2006; **45**: 3835. DOI: 10.1002/anie.200600865.

[33] Chowdhury I, Prasher R, Lofgreen K, Chrysler G, Narasimhan S, Mahajan R, Koester D, Alley R, Venkatasubramanian R. On-chip cooling by superlattice-based thin-film thermoelectrics. Nat. Nanotechnol. 2009; **4**: 235. DOI: 10.1038/NNANO.2008.417.

[34] Nielsch K, Bachmann J, Kimling J, Böttner H. Thermoelectric Nanostructures: From Physical Model Systems towards Nanograined Composites. Adv. Energy Mater. 2011; **1**: 713–731. DOI: 10.1002/aenm.201100207

[35] http://www.electronics-cooling.com/2011/09/thin-film-thermoelectrics-today-and- tomorrow/

[36] Heremans JP, Dresselhaus MS, Bell LE, Morelli DT. When thermoelectrics reached the nanoscale. Nat. Nanotechnol. 2013; **8**: 471. DOI: 10.1038/nnano.2013.129.

[37] http://spectrum.ieee.org/computing/hardware/intelled-team-demonstrates-first-chipscale-thermoelectric-refrigerator

[38] Le PH, Liao C-N, Luo CW, Lin J-Y, Leu J. Thermoelectric properties of bismuth-selenide films with controlled morphology and texture grown using pulsed laser deposition. Appl. Surf. Sci. 2013; 285P: 657–663. DOI: 10.1016/j.apsusc.2013.08.107

[39] Le PH, Luo CW. Thermoelectric and Topological Insulator Bismuth Chalcogenide Thin Films Grown Using Pulsed Laser Deposition. INTECH, UK. 2016; 55–84 DOI: 10.5772/65898

[40] Le PH, Liao C-N, Luo CW, Leu J. Thermoelectric properties of nanostructured bismuth–telluride thin films grown using pulsed laser deposition. J. Alloy. Compd. 2014; **615**: 546–552. DOI: 10.1016/j.jallcom.2014.07.018

[41] Chang H-C, Chen T-H, Whang W-T, Chen C-H. Superassembling of Bi_2Te_3 hierarchical nanostructures for enhanced thermoelectric performance. J. Mater. Chem. *A* 2015; **3**: 10459–10465. DOI: 10.1039/c5ta00911a.

[42] Chen T-H, Lin P-Y, Chang H-C, Chen C-H. Enhanced thermoelectricity of three- dimensionally mesostructured $Bi_xSb_{2-x}Te_3$ nanoassemblies: from micro-scaled open gaps to isolated sealed mesopores, Nanoscale. 2017; **9**: 3283. DOI: 10.1039/c7nr00132k.

[43] Li Bassi A, Bailini A, Casari CS, Donati F, Mantegazza A, Passoni M, Russo V, Bottani CE. Thermoelectric properties of Bi–Te films with controlled structure and morphology. J. Appl. Phys. 2009; **105**: 124307. DOI: 10.1063/1.3147870

[44] Makala RS, Jagannadham K, Sales BC. Pulsed laser deposition of Bi_2Te_3-based thermoelectric thin films. J. Appl. Phys. 2003; **94**: 3907. DOI: 10.1063/1.1600524.

[45] Goncalves LM, Couto C, Alpuim P, Rolo AG, Völklein F, Correia JH. Optimization of thermoelectric properties on Bi_2Te_3 thin films deposited by thermal co-evaporation. Thin Solid Films. 2010; **518**: 2816. DOI: 10.1016/j.tsf.2009.08.038.

[46] Noro H, Sato K, Kagechika H. The thermoelectric properties and crystallography of Bi-Sb-Te-Se thin films grown by ion beam sputtering. J. Appl. Phys. 1993; **73**: 1252–1260. DOI: 10.1063/1.353266

[47] Navrátil J, Horák J, Plecháček T, Kamba S, Lošt'ák P, Dyck JS, Chen W, Uher C. Conduction band splitting and

transport properties of Bi_2Se_3. J. Solid State Chem. 2004; **177**: 1704. DOI: 10.1016/j.jssc.2003.12.031.

[48] Hor Y, Richardella A, Roushan P, Xia Y, Checkelsky J, Yazdani A, Hasan M, Ong N, Cava R. p-type Bi_2Se_3 for topological insulator and low-temperature thermoelectric applications. Phys. Rev. B. 2009; **79**: 195208. DOI: 10.1103/PhysRevB.79.195208.

[49] Mzerd A, Sayah D, Tedenac JC, Boyer A. Crystal growth and sticking coefficient of Bi_2Te_3 thin films on Si(1 1 1) substrate. J. Mater. Sci. Lett. 1995; **14**: 194. DOI: 10.1007/BF00318254

[50] Peranio N, Winkler M, Bessas D, Aabdin Z, König J, Böttner H, Hermann RP, Eibl O. Room-temperature MBE deposition, thermoelectric properties, and advanced structural characterization of binary Bi_2Te_3 and Sb_2Te_3 thin films. J. Alloys Compd. 2012; **521**: 163. DOI: 10.1016/j.jallcom.2012.01.108.

[51] Chang H-C, Chen C-H. Self-assembled bismuth telluride films with well-aligned zero- to three-dimensional nanoblocks for thermoelectric applications. CrystEngComm. 2011; **13**: 5956. DOI: 10.1039/c1ce05350g

[52] Deng Y, Liang H-M, Wang Y, Zhang Z-W, Tan M, Cui J-L. Growth and transport properties of oriented bismuth telluride films. J. Alloys Compd. 2011; **509**: 5683–4587. DOI:10.1016/j.jallcom.2011.02.123

[53] Tuyen LTC, Le PH, Luo CW, Leu J. Thermoelectric properties of nanocrystalline Bi_3Se_2Te thin films grown using pulsed laser deposition. J. Alloy. Compd. 2016; 673: 107–114. DOI: 10.1016/j.jallcom.2016.03.006

[54] Sun G, Qin X, Li D, Zhang J, Ren B, Zou T, Xin H, Paschen SB, Yan X. Enhanced thermoelectric performance of n-type Bi_2Se_3 doped with Cu. J. Alloys Compd. 2015; **639**: 9–14. DOI: 10.1016/j.jallcom.2015.03.124.

[55] Liu W, Lukas KC, McEnaney K, Lee S, Zhang Q, Opeil CP, Chen G, Ren Z. Studies on the Bi_2Te_3–Bi_2Se_3–Bi_2S_3 system for mid-temperature thermoelectric energy conversion. Energy Environ. Sci. 2013; **6**: 552–560. DOI: 10.1039/c2ee23549h

[56] Wang S, Tan G, Xie W, Zheng G, Li H, Yang J, Tang X. Enhanced thermoelectric properties of $Bi_2(Te_{1-x}Se_x)_3$- based compounds as n-type legs for low-temperature power generation. J. Mater. Chem. 2012; **22**: 20943. DOI: 10.1039/c2jm34608

[57] Soni A, Yanyuan Z, Ligen Y, Aik MKK, Dresselhaus MS, Xiong Q. Enhanced thermoelectric properties of solution grown $Bi_2Te_{3-x}Se_x$ nanoplatelet composites. Nano Lett. 2012; **12**: 1203–1209. DOI: 10.1021/nl2034859

[58] Bailini A, Donati F, Zamboni M, Russo V, Passoni M, Casari CS, Li Bassi A, Bottani CE. Pulsed laser deposition of Bi_2Te_3 thermoelectric films. Appl. Surf. Sci. 2007; **254**: 1249. DOI: 10.1016/j.apsusc.2007.09.039.

[59] Obara H, Higomo S, Ohta M, Yamamoto A, Ueno K, Iida T. Thermoelectric Properties of Bi_2Te_3-Based Thin Films with Fine Grains Fabricated by Pulsed Laser Deposition. Jpn. J. Appl. Phys. 2009; **48**: 085506. DOI: 10.1143/JJAP.48.085506.

[60] Takashiri M, Takiishi M, Tanaka S, Miyazaki K, Tsukamoto H. Thermoelectric properties of n-type nanocrystalline bismuth-telluride-based thin films deposited by flash evaporation. J. Appl. Phys. 2007; **101**: 074301. DOI: 10.1063/1.2717867.

[61] Liao C-N, Su X-W, Liou K-M, Chu H-S. Electrical and thermal transport properties of electrically

stressed Bi–Sb–Te nanocrystalline thin film. Thin Solid Films 2011; **519**: 4394. DOI: 10.1016/j.tsf.2011.02.066.

[62] Liao C-N., Wang Y-C., Chu H-S. Thermal transport properties of nanocrystalline Bi–Sb–Te thin films prepared by sputter deposition. J. Appl. Phys. 2008; 104: 104312. DOI: 10.1063/1.3026728

[63] Wang G, Endicott L, Uher C. Recent advances in the growth of Bi–Sb–Te–Se thin films. Sci. Adv. Mater. 2011; 3: 539–560. DOI: 10.1166/sam.2011.1182

[64] Cahill DG. Thermal conductivity measurement from 30 to 750 K: the 3ω method. Rev. Sci. Instrum. 1990; **61**: 802. DOI: 10.1063/1.1141498.

[65] Takashiri M, Miyazaki K, Tanaka S, Kurosaki J, Nagai D, Tsukamoto H. Effect of grain size on thermoelectric properties of n-type nanocrystalline bismuth-telluride based thin films. J. Appl. Phys. 2008; **104**: 084302. DOI: 10.1063/1.2990774.

[66] Takashiri M, Tanaka S, Miyazaki K, Tsukamoto H. Cross-plane thermal conductivity of highly oriented nanocrystalline bismuth antimony telluride thin films. J. Alloys Compd. 2010; 490: L44–L47. DOI: 10.1016/j.jallcom.2009.10.117

[67] Venkatasubramanian R. Lattice thermal conductivity reduction and phonon localizationlike behavior in superlattice structures. Phys. Rev. B. 2000; 61: 3091–3097. DOI: 10.1103/PhysRevB.61.3091

[68] Peranio N, Eibl O, Nurnus J. Structural and thermoelectric properties of epitaxially grown Bi_2Te_3 thin films and superlattices. J. Appl. Phys. 2006; 100: 114306. DOI: 10.1063/1.2375016.

[69] N. F. Hinsche, B. Yu. Yavorsky, M. Gradhand, M. Czerner, M. Winkler, J. König, H. Böttner, I. Mertig, and P. Zahn. Thermoelectric Properties of Bi_2Te_3/ Sb_2Te_3 Superlattices. Phys. Rev. B. 2012; **86**: 085323. DOI: 10.1103/PhysRevB.86.085323

[70] Walachová J, Zeipl R, Zelinka J, Malina V, Pavelka M, Jelínek M, Studnička V, Lošt'ák P. High room-temperature figure of merit of thin layers prepared by laser ablation from Bi_2Te_3 target. Appl. Phys. Lett. 2005; **87**: 081902. DOI: 10.1063/1.2001755.

[71] Chang H-C., Chen C-H., Kuo Y-K. Great enhancements in the thermoelectric power factor of BiSbTe nanostructured films with well-ordered interfaces. Nanoscale. 2013; **5**: 7017–7025. DOI: 10.1039/c3nr01499a

Section 2

Properties

Chapter 4

Characterization of Nanocrystalline Cores for EMI Suppression in Cables

Adrian Suarez, Jorge Victoria, Jose Torres, Pedro A. Martinez, Andrea Amaro and Julio Martos

Abstract

Electromagnetic interferences (EMI) can cause different kinds of problems in digital and analog systems, leading to malfunctions, system reboots, or even permanent damage to the system if this is not adequately designed or protected. Nowadays, most electronic products are connected to the main power network or are designed to be interconnected with others through cables. These cable interconnections are becoming more difficult due to the rigid restrictions related to the accomplishment of electromagnetic compatibility (EMC) compliance. When the cables of a system represent an EMI source, it cannot pass the conducted or radiated emissions test. A widely used technique to reduce these problems is applying an EMI suppressor such as a sleeve core. This EMI suppressor provides selective attenuation of undesired interference components that the designer may wish to suppress, and it does not significantly affect the intended signal. This contribution focuses on analyzing different nanocrystalline (NC) EMI suppressors' performance intended for attenuating interferences in cables. Some NC novel samples are characterized and compare to MnZn and NiZn cores to determine this novel material's effectiveness compared to the conventional ceramic solutions by analyzing samples with different dimensions.

Keywords: electromagnetic interference (EMI) suppressors, electromagnetic compatibility (EMC), nanocrystalline (NC), cable filtering, relative permeability, impedance, insertion loss

1. Introduction

Electromagnetic interference (EMI) can be defined as electromagnetic signals that unintentionally disturb an electrical or electronic system's normal operation. These perturbances can affect the electrical or magnetic magnitudes (voltage, current or electromagnetic field) of its circuits.

The problem of interferences is an issue that design engineers continually face [1]. Electromagnetic interferences can cause different kinds of problems in digital and analog systems, leading to malfunctions, system reboots, or even permanent damage to the system if the system is not adequately designed or protected [2]. The security of an electronic system in which coexist devices that produce electromagnetic interference and small signal circuits that can sensitive to these disturbances,

IntechOpen

depends on the compatibility of the signal levels used.. Thereby, it is convenient to comply with specific design and installation rules that allow making the disturbance levels generated by the interferences source elements compatible with the signal levels used by the possible victim elements or elements sensitive to such interferences [3].

Some standards establish the maximum limits of interferences to ensure that the equipment is compatible and does not interfere. Thus, electromagnetic compatibility (EMC) is the ability of a system to operate satisfactorily in its electromagnetic environment without introducing disturbances above the normalized limits in that environment and withstanding those produced by other equipment. Electromagnetic compatibility is regulated by standards that require compliance with the limits of electromagnetic interferences in electronic systems by studying all the phenomena of generation, propagation, and susceptibility to EMI. Thereby, it is necessary to carry out measurements to certify that this equipment complies with regulations to meet these electronic equipment requirements.

When analyzing an EMI problem, the following elements should be identified (**Figure 1**): the source of interferences, the path of propagation, and receivers affected by the interferences. Based on this concept, when a designer faces an EMI problem, he/she must analyze the system, identify these three elements, and deal with interferences applying these strategies: eliminate EMI sources, increase the EMI immunity of the victim element and/or decrease the energy transmitted through the propagation path.

EMI can spread through different means or paths, as shown in **Figure 2**, so they can be grouped into:

- Conducted interference: when the propagation path is an electrical conductor that joins the sources with the affected receiver such as power cables, signal cables, metal chassis.

- Radiated interferences: they can be classified as far- or near-field depending on the propagation's wavelength and the distance between the source and victim elements. Radiated far-field interferences are identified when carried out through electromagnetic fields, fulfilling the following condition: propagation distance > wavelength/2π. Radiated near-field interferences are called coupling and can be identified as inductive coupling or capacitive coupling between neighboring conductors, depending on whether the interference is propagated by a magnetic or electric field, respectively.

The most appropriate strategy is to consider electromagnetic compatibility during the system design stage. If EMC is ignored until the problem arises during the first functional tests or product certification, the solutions usually result in a higher cost [4]. The possibility of applying specific techniques for the elimination of interferences is reduced as a system is developed. At the same time, the cost of EMI reduction increases [5]. However, it is not always possible to predict EMI problems during the design stage because it is complicated to emulate the real environment in

Figure 1.
Main elements in electromagnetic interference phenomena.

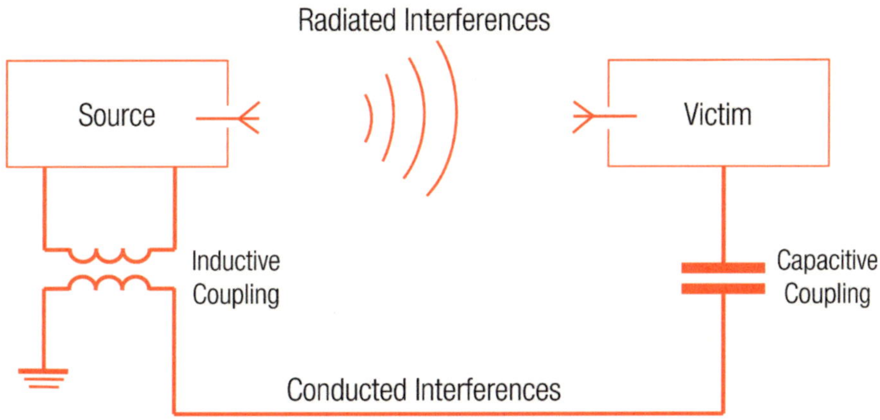

Figure 2.
Kind of propagation and coupling of electromagnetic interference.

which the system will work. Another possibility is that the designed system complies with the standards, but the problems appear when interconnected with other equipment or facilities. When this situation occurs, the right approach is to suppress EMI at its source whenever possible, rather than increasing immunity through victim circuit protections. This technique works best since a single EMI source can find multiple spread paths and affect different victims. If it is not possible to act directly on the source, it is recommended to focus on the EMI propagation path or, finally, on the affected receiver.

Detection and correct characterization of the EMI is an essential factor in designing a suitable solution. Thus, it is essential to perform EMI measurements using different instrumentation, measuring probes, and antennas to detect the electromagnetic fields that can provide information to the designer from undesired signals. These measurements make it possible to detect the disturbances' magnitude and localize their frequency range in order to select the most optimal solution.

When the cables represent the EMI source, it could not pass the conducted or radiated emissions test. A widely used technique to reduce these problems is applying an EMI suppressor such as a sleeve core [5].

This contribution focuses on analyzing different nanocrystalline (NC) EMI suppressors' performance intended for attenuating interferences in cables. Firstly, some applications of this kind of EMC components are described in section 2, while the description of the manufacturing process and main features of NC material are explained in section 3. The characterization methods employed to determine the NC samples' effectiveness from the standpoint of the impedance and insertion loss they can provide are shown in section 4. Subsequently, in section 5, NC sleeve cores' performance is discussed and compared to conventional ceramic samples. Finally, the main conclusions are summarized in section 6.

2. Applications of EMI suppressor sleeve cores

The Magnetic Field (H) is associated with electrodynamic phenomena and appears whenever there are electric currents. The H field can produce effects capable of seriously disturbing the operation of an electronic circuit. Whenever current flows in a circuit, this current creates a magnetic field in that circuit, which will vary as the current varies. Consequently, in any circuit that carries an alternating current, variations of magnetic flux occur. According to Lenz's law, an electromotive force will be induced by the field variation. Therefore, if the current is constant, there will be no induced electromotive force.

Considering that the flux density (B) is proportional to the product of the permeability of the medium and the incident H field, B is the result of the action of H in a magnetic circuit, and its intensity will be higher or lower depending on the permeability of the matter (μ_r). For the shielding of conductors against EMI, the most common is to use ferromagnetic materials since they present a permeability much higher than that of vacuum (μ_0).

When introducing the sample, the external field deforms considerably, being, at each point, the resultant of the initial magnetic field and the field created by the orientation of the magnetic domains. As shown in **Figure 3**, the material concentrates the field lines and regions outside and close to the material, reducing the emitted field.

As explained above, unexpected EMI sources in cables can appear in our system when connected to another device. One of the most used techniques for reducing cables' interferences is applying an EMI suppressor such as sleeve cores to them. This EMI suppressor provides selective attenuation of undesired interference components that the designer may wish to suppress and it does not affect the intended signal. Thereby, this component is widely used to filter EMI in power cables to reduce high-frequency oscillations generated by switching transients or parasitic resonances within a circuit, and EMI in peripheral cables of electronic devices such as multiconductor USB or video cables.

From the standpoint of the magnetic properties, a sleeve core is defined by the relative permeability since it is the main parameter that describes the performance of a specific magnetic material to concentrate the magnetic flux in the core. This parameter is generally expressed through its complex form represented by the real component (μ_r') that quantifies the real or inductive part and the imaginary or resistive component (μ_r") that is related to the material ability to absorb the electromagnetic interferences [6, 7].

The presence of noise current in a conductor generates an undesired magnetic field around it, resulting in EMI problems. The effectiveness of a sleeve core to reduce EMI in cables is defined by its capability to increase the flux density of a certain field strength created around a conductor. Thereby, noise current generates a magnetic field which is concentrated into magnetic flux inside the ferrite by the core's magnetic permeability (μ_r'). This magnetic field inside the ferrite is reduced by the ferrite's magnetic loss (μ_r"), converting it into heat energy. As a result of these two filtering mechanisms the flowing noise current in the conductors is reduced.

Currents that flow in cables (with two or more conductors) can be divided into differential mode (DM) and common mode (CM) depending on the directions of propagation. Although DM currents are usually significantly higher than CM currents, one of the most common EMI radiated problems is originated by CM currents

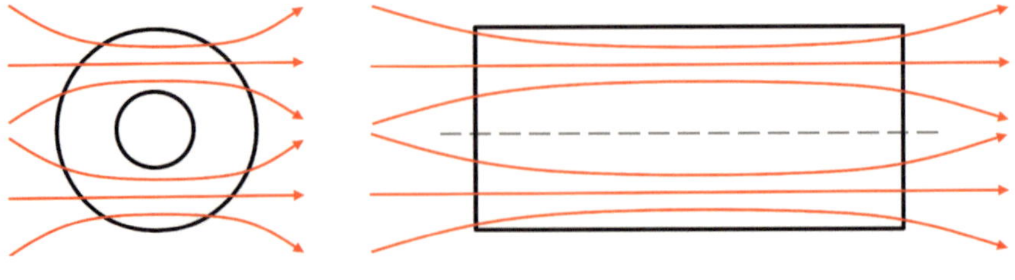

Figure 3.
Management of H field through introducing a sleeve core.

flowing through the cables of the system [8]. CM currents have a much greater interfering potential, despite not having a high value. This fact is due to only a few microamps are required to flow through a cable to fail radiated emission requirements [5, 9]. The use of sleeve cores is an efficient solution to filter the CM currents in cables because, if a pair of adjacent conductors is considered, when the cable ferrite is placed over both signal and ground wires, the CM noise is reduced. As shown in **Figure 4**, the CM currents in both wires flow in the same direction, so the two magnetic fluxes in the cable ferrite are added together, and the filtering action occurs in the sleeve core. The intended (DM) current is not affected by the presence of the cable ferrite because the DM current travels in opposite directions and is transmitted through the signal and returns. Thus, the current of the two conductors is opposing, meaning they cancel out and the cable ferrite has no effect [10, 11]. In the case of wanting to filter the DM currents, it would be necessary to use a sleeve core in each of the cable's conductors.

An external power supply (**Figure 5**) can be considered a specific example of the application of sleeve cores to reduce EMI in terms of both radiated and conducted emissions. Within the conducted emission range (150 kHz – 30 MHz), the conductors of the system are generally too short to be considered an EMI antenna source since the impedance of possible parasitic inductors is low, and the impedance of parasitic capacitors is typically high. Nevertheless, in the radiated emissions range (from 30 MHz), the parasitic associated with conductors and power line EMI filters can be significant if conductors are long enough to be considered an unintended antenna [12, 13]. External power supplies typically incorporate discrete inductors, capacitors in the AC input circuitry to implement common mode, and differential mode filters before the input bridge and the switching stage. This filtering stage's main objective is to attenuate the interferences that can be conducted out from the power supply to the AC input power lines. Accordingly, the internal PCB is designed to hold these filtering components in order to pass regulatory safety and EMC testing. When these techniques are considered, a power supply design may meet conducted and radiated emission requirements when tested in isolation. Nevertheless, when the power supply is added to a complete system, the system may fail emissions testing due to the interferences emitted from the system load to the

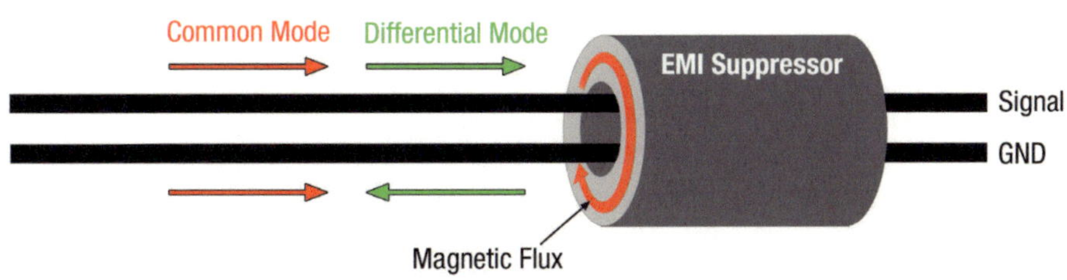

Figure 4.
Diagram of CM and DM currents passing through a cable ferrite with two adjacent conductors (signal and return paths).

Figure 5.
Diagram of an external power supply elements.

designed power supply through the DC output cable back. One of the most common solutions to solve this EMI problem is integrating a sleeve core that reduces the undesired interferences without affecting the DC intended signal.

The advantage of using this EMI solution is that it does not involve redesign the electronics and, generally, the mechanical redesign. This is an important advantage because determining in the testing stage, which is the EMI source, may not be straightforward. However, the use of a sleeve core involves adding an extra component whose drawbacks result in increasing the product's size and weight besides the cost of the filtering component and its installation. Therefore, this is an effective solution to attenuate EMI emissions in cables when it is not possible to solve the problem through a system redesign, but it is essential to strike a balance between performance and other factors such as weight, dimensions, and cost [14].

Sleeve cores are manufactured with magnetic material that allows them to control interferences in a certain frequency range with a specific ratio. The values of these two parameters mainly depend on the EMI suppressor intrinsic composition and internal structure. The most used sleeve ferrite cores are based on ceramics or polycrystalline materials because they contain metal oxides, such as manganese or zinc oxide [15]. Thereby, MnZn and NiZn are the most popular EMI suppressor solution due to their heat resistant, hardness, and high resistance to pressure. One of the ceramics' main advantages is the possibility of manufacturing samples with many different shapes able to provide a significant performance [16, 17]. The starting material of ceramics is iron oxide Fe_2O_3 mixed with one or more divalent transition metals, such as manganese, zinc, nickel, cobalt, or magnesium [18]. Nanocrystalline sleeve core represents an innovative and increasingly used solution for EMI suppression in cables. This solution has demonstrated excellent suitability to reduce interferences from the low-frequency region to the mid-frequency range [19, 20]. In this sense, some researchers have investigated the use of NC structure compositions to make EMC components because this kind of core can reduce its volume by 50–80% and yield greater magnetic properties and insertion losses than conventional ceramic components [21–24].

Consequently, some NC novel samples are characterized and compare to MnZn and NiZn cores to determine this novel material's effectiveness compared to the conventional ceramic solutions by analyzing samples with different dimensions.

3. Nanocrystalline core description

The manufacturing procedure of ceramic materials (**Figure 6**) is based on, firstly, mixing raw materials into the desired proportions. Next, it is then pre-calcined to form the ferrite. The pre-sintered material is then milled to obtain a specific particle size. Subsequently, the granulated material is shaped by a pressing technique to obtain the final form. Finally, the resultant core is sintered, promoting any unreacted oxides to be formed into ferrite and protected with epoxy [25, 26]. The manufacturing procedure and the material mix are essential to define a ceramic core's magnetic properties. Thereby, MnZn materials can provide a significant performance for EMI suppression applications, covering the range of frequency from hundreds of kHz to some MHz. In contrast, NiZn materials are intended for a higher frequency operation than MnZn, covering from tens of MHz to several hundreds of MHz [16, 19, 20, 27].

Figure 7 shows two micrographs of the samples obtained using scanning electron microscopy (SEM). In these photographs, it is possible to observe the internal structure and the grain size of MnZn (a) and NiZn (b) ceramic materials.

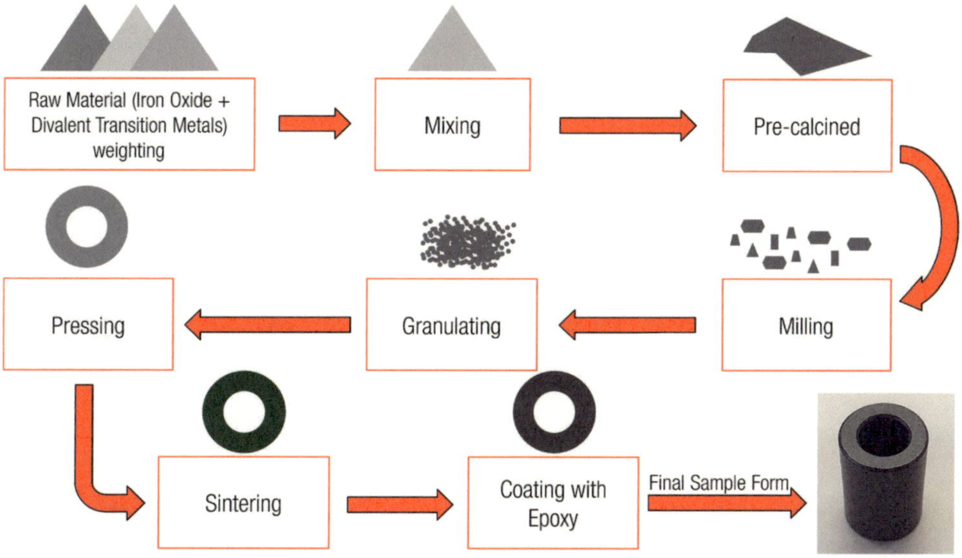

Figure 6.
Diagram of the manufacturing procedure of ceramic cores.

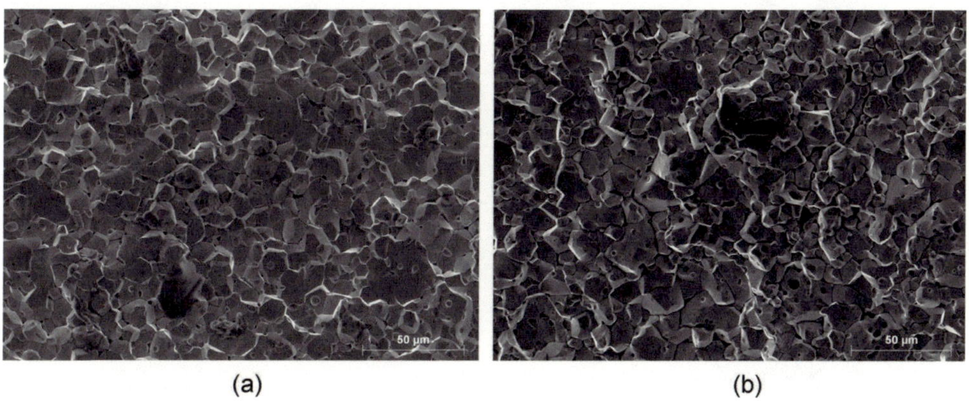

Figure 7.
SEM photographs of ceramics core materials: (a) MnZn material composition; (b) NiZn material composition.

The results presented in [19, 20] highlighted the great suitability of NC EMI suppressors to filter electromagnetic interference throughout the frequency band from 100 kHz to 100 MHz. Furthermore, data obtained from its magnetic properties indicate that this EMC solution could also provide good performance in terms of EMI suppression at higher frequencies. The advantages of iron-based nanocrystalline materials lie in the high values of relative permeability, the reduction of the magnetic components' volume, and the stable operation up to high-temperatures. These properties are mainly defined by the manufacturing procedure. The manufacturing procedure of NC samples (**Figure 8**) is quite different from the used for ceramic production since it is formed by a continuous laminar structure that is wound to form the final core. The material is a two-phase structure consisting of an ultra-fine grain phase of FeSi embedded in an amorphous ribbon of 7–25 micrometers in thickness. Firstly, the base material is molten by heating it at 1300 °C and depositing it on a water-cooled wheel that reduces the temperature of the material to 20 °C. Next, the resulting amorphous metal ribbon is exposed to an annealing

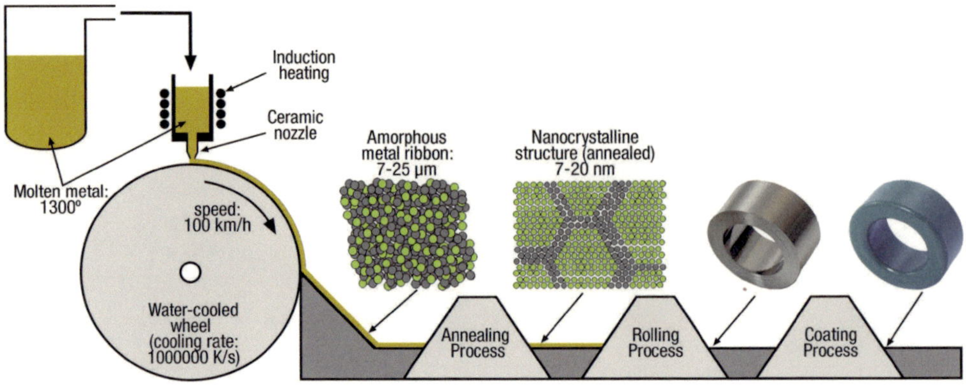

Figure 8.
Diagram of the manufacturing procedure of nanocrystalline cores..

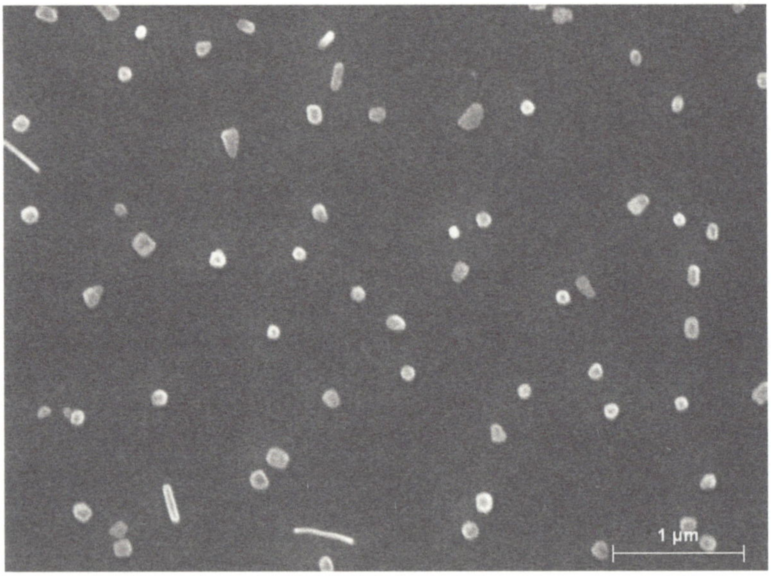

Figure 9.
SEM photograph of a nanocrystalline core material.

process under the presence of transversal and/or longitudinal magnetic fields. This treatment modifies the magnetic properties and the amorphous structure forms ultrafine crystals with a typical size of 7–20 nm, obtaining the nanocrystalline material. The last stage of this procedure corresponds to applying a protective coating or a plastic housing that protects the obtained cores due to the brittle nature of the tape [11, 18, 28].

Figure 9 shows a SEM photograph of a NC sample where it is possible to observe the difference in terms of the grain size if it is compared with the ceramic materials since it is in the order of nanometers.

4. Characterization methods

The evaluation of NC sleeve cores is carried out by analyzing the performance regarding two MnZn and NiZn ceramic cores. Therefore, the evaluation of the three different solutions in terms of EMI suppression from the standpoint of the magnetic

Sample reference	Magnetic material	Outer diameter (OD) (mm)	Inner diameter (ID) (mm)	Height (H) (mm)
S1	NC	15.3	5.5	20.0
S2	MnZn	16.0	8.0	28.5
S3	NiZn	17.5	9.5	28.5
S4	NC	28.3	15.5	30.0
S5	MnZn	26.0	13.0	28.5
S6	NiZn	26.0	13.0	28.5

Table 1.
List of sleeve core samples analyzed.

properties, the impedance, and the insertion loss provided by different samples. One of the cores selected is based on MnZn, a material widely used to reduce EMI in the low-frequency region and the other selected core is made of NiZn that is generally employed to filter EMI from some tens of megahertz. Thus, it is important that the analyzed sleeve cores have a similar volume in order to conclude which solutions is more effective depending on the frequency range selected. Accordingly, different ceramic MnZn and NiZn sleeve core samples with similar dimensions to the NC samples have been selected to be characterized and evaluated, as shown in **Table 1**. Note that two sets of three different materials are analyzed. In the case of the small samples set (S1, S2, and S3), the ceramic samples are longer than the NC one, whereas in the large samples set (S4, S5, and S6), the three cores have similar dimensions.

4.1 Relative permeability

The relative permeability (μ_r) is one of the most important parameters that define the material's ability to absorb electromagnetic interferences. The permeability relates the magnetic flux density of a specific magnetic field in a defined medium. When a sleeve core is placed around a certain cable, it concentrates the magnetic flux. The material's internal properties describe its ability to focus the magnetic flux is represented through the permeability complex parameter. The effectiveness to attenuate EM interferences of a material can be quantified by separating μr into its complex form. The real component is related to the stored energy or inductive part (μ') and the imaginary component that provides the losses or resistive part (μ"). Thereby, the complex relative permeability is expressed by:

$$\mu_r(f) = \mu'(f) - j\mu''(f) \tag{1}$$

The magnitude of the NC material's relative permeability is represented together with MnZn and NiZn permeability traces in **Figure 10** to study the frequency region covered by each material. This graph shows the NC core provides higher permeability than the ceramic materials throughout almost the entire frequency range studied, despite being the material with higher initial permeability. MnZn has an initial permeability (μ_i) of 5000 and it is able to provide a permeability around 3000 up to the 2 MHz, providing a similar value to NC at this frequency point. NC demonstrates the best performance in the mid-frequency region, whereas the NiZn material (μ_i = 620) is more effective in the high-frequency region. The NC material has an initial permeability (μ_i) of 30000 and it provides a significant permeability up to 200 MHz.

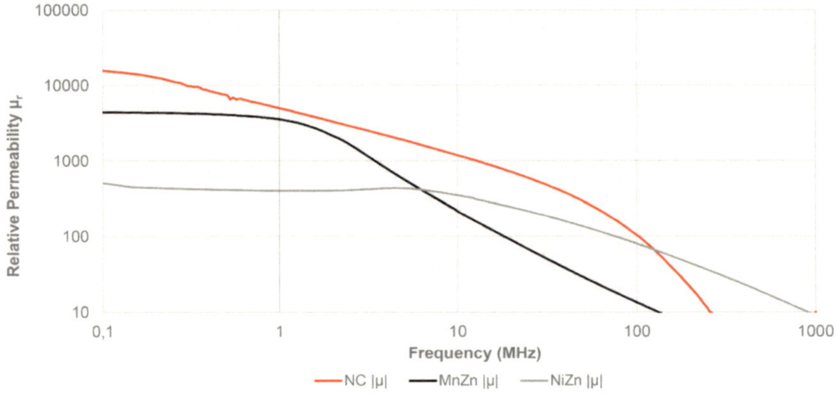

Figure 10.
Relative permeability of NC core compared to MnZn and NiZn compositions.

4.2 Impedance parameter

The permeability parameter is used to describe the core material's behavior; however, manufacturers of EMI Suppressors generally provide their customers the impedance that it introduces in the cable in which it is applied. Typically, the datasheets only specify the impedance at several frequency points or the graph of the magnitude of the impedance in the frequency range where it is more effective. The impedance of a certain sleeve core considers, besides the material permeability, other variables such as the self-inductance defined by the dimensions and the shape. Thereby, sleeve cores are usually defined and classified by specifying the magnitude of the impedance (Z_F), which is obtained from the equivalent component parameters such as resistance (R) and the impedance of the inductive part (X_L). The magnitude of the impedance is given by:

$$|Z_F| = \sqrt{R^2 + (X_L)^2}. \tag{2}$$

The measurement of the impedance carried out in this contribution has been performed by using the E5061B Vector Network Analyzer (Keysight) connected to the Terminal Adapter 16201A (Keysight) and the Spring Clip Fixture 16092A (Keysight), as shown in **Figure 11**. These fixtures are internally compensated by an impedance standard calibration method to consider the electrical length path and the impedance variations caused by parasitic elements.

4.3 Insertion loss parameter

Another kind of EMC component datasheets, such as common-mode-chokes, show the attenuation ratio or insertion loss in terms of decibels (dB) that are able to provide. In the case of sleeve cores, it is also possible to determine the insertion loss that it introduces when applied in a cable. The insertion loss that a sleeve ferrite core is able to yield is strongly dependent on the impedance of the system in which it is placed, besides its impedance response depending on the frequency. Subsequently, these components are more effective against EMI when the source and load systems' impedance is low. The equivalent circuit approach to determine the insertion loss parameter of a specific sleeve core requires considering the source impedance (Z_A) and the load impedance (Z_B) of the system with electromagnetic interference

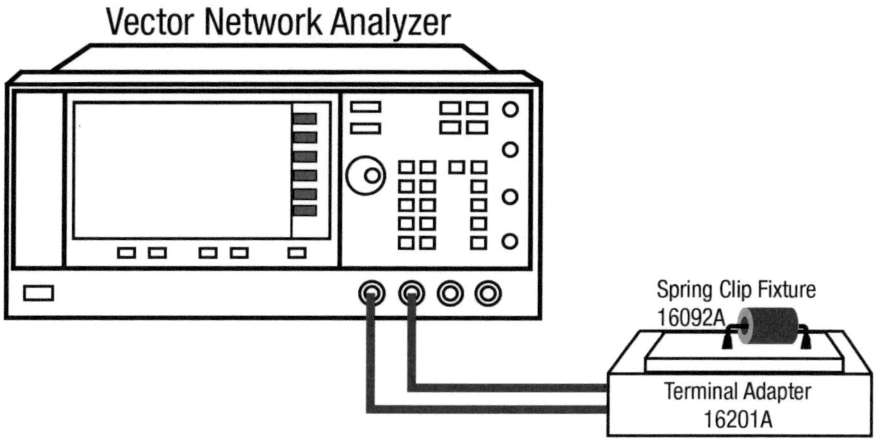

Figure 11.
Setup for measuring impedance of sleeve core samples.

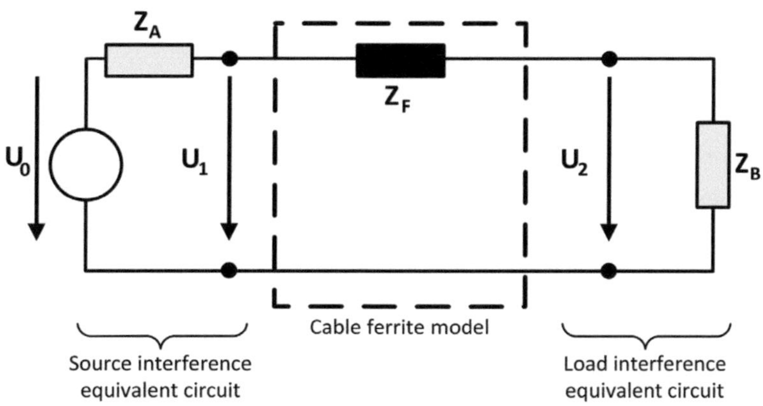

Figure 12.
Schematic of source and load equivalents circuits used to determine the insertion loss parameter of a sleeve core when introduced into a system.

problems. The impedance introduced by the sleeve core (Z_F) in that system is introduced in the path that connects both systems. The equivalent circuit diagram employed to determine this impedance relation and analyze the effect of introducing a sleeve core into a certain system is shown in **Figure 12** [29].

According to this diagram, when the system impedance is known, the insertion loss (A) in terms of decibels can be calculated through the Eq. (3) considering the impedance of the sleeve core (Z_F):

$$A(dB) = 20\log\left(\frac{Z_A + Z_F + Z_B}{Z_A + Z_B}\right) \tag{3}$$

5. Results and discussion

The results presented in this section correspond to the analysis of the performance provided by NC samples compared to ceramic solutions. This comparison is carried out by evaluating both the impedance and insertion loss parameters described in the previous section. Two sets of three sleeve cores are analyzing to study the performance provided by samples intended for thin cables (samples S1, S2, and S3) and those with larger diameters (samples S4, S5, and S6). These results make it possible to observe the performance of each EMI suppression solution. It allows the system designer to select the best component to solve EMI problems depending on the frequency range where it is located.

Firstly, the impedance measured of the two sets of samples is shown in **Figures 13** and **14**. **Figure 13** shows the response of the three small samples and it can be observed that the MnZn sleeve core is able to provide the best performance in the low-frequency, achieving its maximum value at 1.5 MHz (132.08 Ω). NiZn sleeve core reaches the maximum impedance value at 50.1 MHz (145.63 Ω). This material represents an interesting solution to reduce EMI in the mid and high-frequency regions, whereas it does not provide a valuable impedance in the low-region. NC sample offers the highest impedance values in the mid-frequency region (from 4.1 MHz to 95.6 MHz), reaching the maximum impedance value at 34.9 MHz (162.04 Ω).

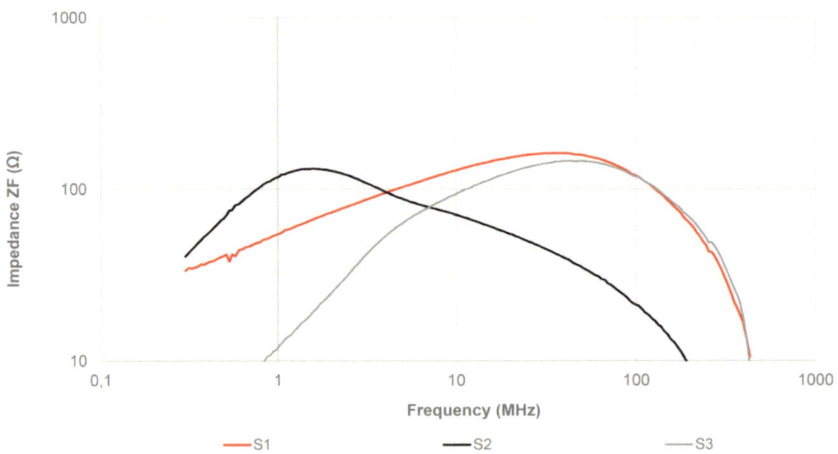

Figure 13.
Magnitude impedance of the NC (S1), MnZn (S2), and NiZn (S3) samples.

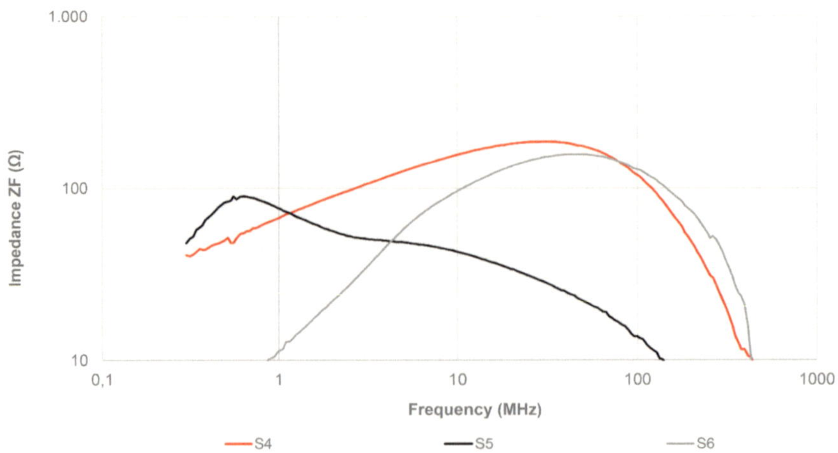

Figure 14.
Magnitude impedance of the NC (S4), MnZn (S5), and NiZn (S6) samples.

Nevertheless, in contrast to ceramic materials, NC sample is able to provide a significant response in both the low and high-frequency region. NC sleeve core shows a better performance than ceramics to reduce EMI emissions in a wideband frequency range.

Figure 14 shows the performance of the large sleeve core samples in terms of the impedance response. It is possible to observe that the MnZn sample is the most effective solution to reduce interferences up to 1.2 MHz. From this frequency value, the NC sample is able to introduce a higher impedance value than the other two solutions, covering the range from 1.2 MHz to 77.7 MHz. NiZn sample provides larger impedance than NC in the high-frequency region. The maximum impedance value offered for the MnZn sample is located at 0.6 MHz (89.55 Ω). In the case of the NC sample, this value is achieved at 33.1 MHz (186.77 Ω), whereas the NiZn sample reaches the highest impedance value at 48.2 MHz (157.49 Ω). It is possible to observe that the frequency ranges where the large sleeve core is most effective are similar to the provided by the small samples. In **Figure 14**, the MnZn sample (S5) shows a less significant performance than the smaller sample response based on the same material (S2). The dimensional effect causes this shift in the resonance frequency (maximum impedance value). Thereby, MnZn material reduces its performance when it is used to manufacture large cores due to its internal structure and electrical features. Nevertheless, it is possible to observe that the dimensional effect does not affect the NC and NiZn material when used to manufacture large EMI suppressor cores. However, despite these aspects, the effectiveness in the different frequency regions is similar to the described for the last set of samples.

The insertion loss results have been obtained by considering a system with an input and output impedance of 50 Ω ($Z_A = Z_B = 50$ Ω). Thereby, the experimental results that are shown in **Figures 15** and **16** can be compared by considering Eq. (3) and the impedance provided by each sleeve core (Z_F). The results obtained in terms of insertion loss correlate with the impedance responses shown previously since MnZn samples provide a higher attenuation ratio in the low-frequency region. Specifically, the S2 sample provides up to −7.25 dB at 1.7 MHz and − 5.43 dB at 0.6 MHz in the case of S4. Thereby, MnZn material represents the best solution when the interferences are located below 4.2 MHz, considering the small sample set (see **Figure 15**). This frequency range is reduced when larger samples are analyzed since the MnZn S4 sample predominant frequency range

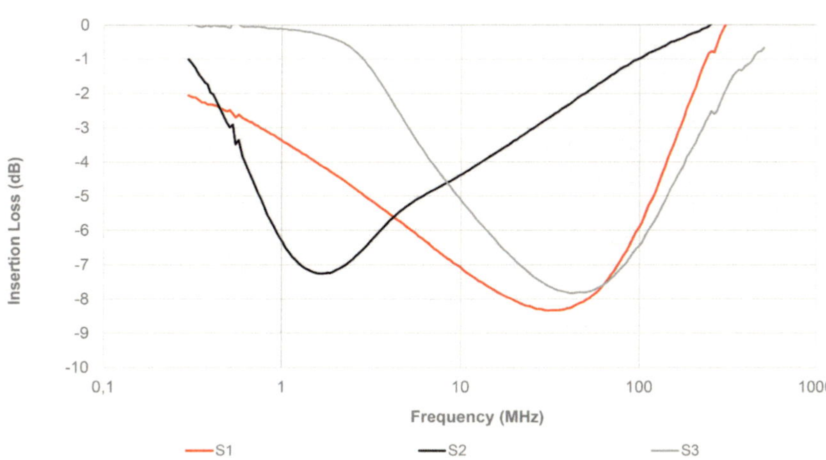

Figure 15.
Insertion loss of the NC (S1), MnZn (S2), and NiZn (S3) samples.

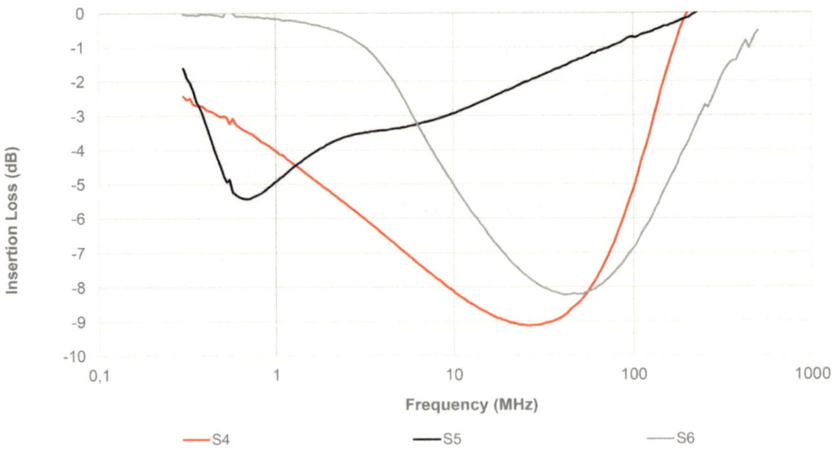

Figure 16.
Insertion loss of the NC (S4), MnZn (S5), and NiZn (S6) samples.

is shorter up to 1.3 MHz if compared to the NC S3 sample (see **Figure 16**). As a result of this, the effectiveness of MnZn material is reduced compared to NC material when the dimensions of the sample are increased, since the S4 sample's frequency range is shorted 3.9 MHz and the maximum attenuation has been reduced 1.82 dB. Considering this behavior of MnZn material when a large core is employed, NC represents an alternative solution to suppress EMI in the low-frequency region when a large sleeve core is needed due to it provides higher attenuation than NiZn in this range. Regarding the mid-frequency region, NC samples (S1 and S4) have a similar response, reaching the maximum values of insertion loss at 33.5 MHz (−8.33 dB) in the case of S1 and 28.2 MHz (−9.11 dB) for S4. NC S1 sample has the predominant response from 4.2 MHz to 60.4 MHz, considering the small sample set (see **Figure 15**) and from 1.3 MHz to 57.2 MHz in the case of the large samples (see **Figure 16**). NiZn S3 and S6 samples are able to offer the best performance in the high-frequency region since NC samples have a resonance frequency lower than the value shown by NiZn cores. This insertion loss difference between NC and NiZn in the high-frequency region is more significant when the large sample cores.

Consequently, MnZn samples are significantly effective in the low-frequency region, but their performance is strongly reduced in the high-frequency region. Contrary to this behavior, NiZn samples show great insertion loss in the high-frequency region, whereas it provides a poor performance in the low-region. However, NC samples show the best performance in the mid-frequency region at the same time that it provides a significant insertion loss in the low-frequency region and a comparable response than the offered by the NiZn samples in the high-frequency region.

Note that these results are related to the impedance of both systems where the cable in which the sleeve core is applied. Therefore, the insertion loss values obtained can be considered when the EMI suppression solution is applied to data or video cables. According to Eq. (3), if these samples were installed in power cables, it could be possible to obtain higher attenuation ratios. For instance, if the sleeve core is installed in a system where $Z_A = Z_B = 5\ \Omega$, the Z_F provided by the sample is more significant than the system impedance. Thereby, if the maximum impedance provided by S1 is considered ($Z_F = 145.63\ \Omega$) it can be able to introduce an insertion loss of −23.84 dB instead of the −8.33 dB obtained for $Z_A = Z_B = 50\ \Omega$.

6. Conclusions

The performance of NC samples has been compared with the effectiveness provided by ceramic cores. Thereby, it has been analyzed the performance of each EMI suppression solution from the standpoint of the magnetic properties, impedance, and insertion loss.

Considering the results presented, it is possible to identify the frequency regions where each material solution is effective to reduce EMI when applied in a certain cable. According to the relative permeability data, the MnZn material analyzed is suitable when the interferences are located in the low-frequency region (from hundreds of kilohertz up to some megahertz). In contrast, NiZn solution is not effective in this frequency region. NiZn samples show an interesting solution to reduce EMI in the mid and high-frequency region since it shows a better response than MnZn up to about 5 MHz. The relative permeability data shows that NC material is able to provide a wideband solution due to it is able to offer a comparable response to MnZn material in the low-frequency region and NiZn material in the high-frequency region. Furthermore, NC shows the highest permeability in the mid-frequency region. The excellent magnetic properties shown by the NC material have been verified from the standpoint of the impedance and the insertion loss that the NC samples can introduce in a certain cable with electromagnetic disturbances. Therefore, MnZn samples show a significant performance to reduce EMI in the low-frequency region in terms of impedance and insertion loss, whereas NiZn is effective against high-frequency interferences.

Consequently, if the EMI disturbances are specifically located in the low or high-frequency region, a ceramic core is able to provide significant effectiveness to reduce them. If the interferences are detected in the mid-region (from 5 MHz to 100 MHz), NiZn material is able to provide better performance than MnZn if only ceramic cores are considered. NC structures usually represent a higher cost than ceramic, so that this solution may not always be considered to solve an EMI problem located in a specific frequency region. This is the reason why a designer could select a ceramic core instead of a NC core to reduce an EMI problem despite the ceramic core could not be the most effective solution. Nevertheless, when the EMI disturbances are distributed in different frequency regions, NC sleeve core shows a better performance than ceramics to reduce EMI emissions in a wideband frequency range.

Author details

Adrian Suarez*, Jorge Victoria, Jose Torres, Pedro A. Martinez, Andrea Amaro and Julio Martos
School of Engineering of the University of Valencia, Valencia, Spain

*Address all correspondence to: adrian.suarez@uv.es

IntechOpen

© 2021 The Author(s). Licensee IntechOpen. This chapter is distributed under the terms of the Creative Commons Attribution License (http://creativecommons.org/licenses/by/3.0), which permits unrestricted use, distribution, and reproduction in any medium, provided the original work is properly cited.

References

[1] González-Vizuete P, Domínguez-Palacios C, Bernal-Méndez J, Martín-Prats MA. Simple Setup for Measuring the Response to Differential Mode Noise of Common Mode Chokes. Electronics. 2020;9. DOI: 10.3390/electronics9030381

[2] Valenzuela R. Novel Applications of Ferrites. Physics Research International. 2012;2012. DOI: 10.1155/2012/591839

[3] Balcells J, Daura F, Esparza R, Pallás R. Interferencias electromagnéticas en sistemas electrónicos. Barcelona: Marcombo; 1992

[4] Crovetti PS, Musolino F. Interference of Spread-Spectrum EMI and Digital Data Links under Narrowband Resonant Coupling. Electronics. 2020;9. DOI: 10.3390/electronics9010060

[5] Ott HW. Electromagnetic Compatibility Engineering. Hoboken: John Wiley & Sons; 2009.

[6] Suarez A, Victoria J, Alcarria A, Torres J. Characterization of electromagnetic noise suppression sheet for aerospace applications. In: Proceedings of the ESA Workshop on Aerospace EMC; 23-25 May 2016; Valencia. New York: IEEE; 2016. p. 1-6

[7] Victoria J, Suarez A, Torres J, Martinez PA, Alcarria A, Martos J, Garcia-Olcina R, Soret J, Muetsch S, Gerfer A. Transmission Attenuation Power Ratio Analysis of Flexible Electromagnetic Absorber Sheets Combined with a Metal Layer. Materials. 2018;11. DOI: 10.3390/ma11091612

[8] Yao J, Li Y, Zhao H, Wang S. Design of CM Inductor Based on Core Loss for Radiated EMI Reduction in Power Converters. In: Proceedings of the 2019 IEEE Applied Power Electronics Conference and Exposition (APEC); 17-21 March 2019; Anaheim. New York: IEEE; 2019. p. 2673-2680

[9] Bondarenko N, Shao P, Orlando A, Koledintseva MY, Beetner DG, Berger P. Prediction of common-mode current reduction using ferrites in systems with cable harnesses. In: Proceedings of the 2012 IEEE International Symposium on Electromagnetic Compatibility; 05-10 August 2012; Pittsburgh. New York: IEEE; 2012. p. 80-84

[10] Kraftmakher Y. Experiments on ferrimagnetism. European Journal of Physics. 2012;34. DOI: 10.1088/0143-0807/34/2/213

[11] Suarez A, Victoria J, Torres J, Martinez PA, Alcarria A, Perez J, Garcia-Olcina R, Soret J, Muetsch S, Gerfer A. Performance Study of Split FerriteCoresDesignedforEMISuppression on Cables. Electronics. 2020;9. DOI: 10.3390/electronics9121992

[12] Suarez A, Victoria J, Torres J, Martinez PA, Martinez V, Molina I, Muetsch S, Garcia-Olcina R, Soret J, Martos J. Determination of Core Size Dependency on the EMI Suppression in Cable Ferrites. In: Proceedings of the 22020 International Symposium on Electromagnetic Compatibility-EMC EUROPE; 23-25 September 2020; Barcelona. New York: IEEE; 2020. p. 1-6

[13] Chiu HJ, Pan TF, Yao CJ, Lo YK. Automatic EMI measurement and filter design system for telecom power supplies. IEEE Trans. Instrum. Meas. 2007;56. DOI: 10.1109/TIM.2007.907948

[14] Williams, T. EMC for Product Designers. Burlington: Elsevier Science & Technology; 2006.

[15] Shaikh SF, Ubaidullah M, Mane RS, Al-Enizi AM. Types, Synthesis methods and applications of ferrites. In: Mane RS, Vijaykumar VJ, editors. Spinel

Ferrite Nanostructures for Energy Storage Devices. Burlington: Elsevier Science & Technology; 2020.

[16] Cullity BD, Graham CD. Introduction to Magnetic Materials. Hoboken: Wiley; 2009.

[17] Vaid K, Chaurasia A, Rathore D, Singhal R, Dwivedi UK. Study of dielectric and electromagnetic shielding behaviour of BaTiO3-CoFe2O4 filled LDPE composite. Polymer Composites. 2020. DOI: 10.1002/pc.25867

[18] Van den Bossche A, Valchev VC. Inductors and Transformers for Power Electronics. Boca Raton: CRC Press; 2005.

[19] Suarez A, Victoria J, Alcarria A, Torres J, Martinez PA, Martos J, Soret J, Garcia-Olcina R, Muetsch S. Characterization of Different Cable Ferrite Materials to Reduce the Electromagnetic Noise in the 2-150 kHz Frequency Range. Materials. 2018;11. DOI: 10.3390/ma11020174

[20] Suarez A, Victoria J, Torres J, Martinez PA, Alcarria A, Perez J, Garcia-Olcina R, Soret J, Muetsch S, Gerfer A. Effectiveness Assessment of a Nanocrystalline Sleeve Ferrite Core Compared with Ceramic Cores for Reducing Conducted EMI. Electronics. 2019;8. DOI: 10.3390/electronics8070800

[21] Herzer G, Vazquez M, Knobel M, Zhukov A, Reininger T, Davies HA, Sanchez JS. Round table discussion: Present and future applications of nanocrystalline magnetic materials. J. Magn. Magn. Mater. 2005;294. DOI: 10.1016/j.jmmm.2005.03.042

[22] Thierry W, Thierry S, Benoît V, Dominique G. Strong volume reduction of common mode choke for RFI filters with the help of nanocrystalline cores design and experiments. J. Magn. Magn. Mater. 2006;304. DOI: 10.1016/j.jmmm.2006.03.014

[23] Herzer G. Modern soft magnets: Amorphous and nanocrystalline materials. Acta Materialia. 2013;61. DOI: 10.1016/j.actamat.2012.10.040

[24] Liu Y, Han Y, Liu S, Lin F. Pulse Magnetic Properties Measurement and Characterization of Fe-Based Nanocrystalline Cores for High-Voltage Pulse Magnetics Applications. IEEE Trans. Power Electron. 2015;30. DOI: 10.1109/TPEL.2014.2386916

[25] Zaspalis V, Tsakaloudi V, Papazoglou E, Kolenbrander M, Guenther R, Van Der Valk P. Development of a new MnZn-ferrite soft magnetic material for high temperature power applications. Journal of Electroceramics. 2004;13. DOI: 10.1007/s10832-004-5162-3

[26] Miszczyk A. Protective and suppressing electromagnetic interference properties of epoxy coatings containing nano-sized NiZn ferrites. Frontiers in Materials. 2020;7. DOI: 10.3389/fmats.2020.00183

[27] Kącki M, Rylko MS, Hayes JG, Sullivan CR. Magnetic material selection for EMI filters. In: Proceedings of the 2017 IEEE Energy Conversion Congress and Exposition (ECCE); 01-05 October 2017; Cincinnati. New York: IEEE; 2017. p. 2350-2356

[28] Kazimierczuk MK. High-Frequency Magnetic Components. Hoboken: Wiley; 2009.

[29] Brander T, Gerfer A, Rall B, Zenkner H. Trilogy of Magnetics: Design Guide for EMI Filter Design, SMP & RF Circuits. Künzelsau: Swiridoff Verlag; 2010.

Chapter 5

Fullerenes and Nanodiamonds for Medical Drug Delivery

Basma H. Al-Tamimi and Saad B.H. Farid

Abstract

Carbon is a chemical element has the ability of forming long carbonic chain. Due to its special electronic structure, each carbon atom can be linked with another carbon atom or with another element via single, double or triple covalent bonds. The special electronic structure of carbon atom affecting on its properties also affecting on its ability of existing in different forms called allotropes. During few last decades, new carbon-based nanomaterials have been described including fullerene, carbon nanotube, graphene and nanodiamond. These new allotropes attracted the interest of science and industry and became as a new and important class of materials due to its outstanding features which candidate for numerous applications. In parallel with new developments in nanomedicine especially in drug delivery field, the targeted delivery systems became an important to overcome the limitations of the old fashion systems. So, it become very important to translate this idea into reality. Fullerene and nanodiamond have a unique combination of structure, morphology and biological properties that make them as a powerful tools for targeted delivery system. So, this chapter will focus on two major aspects: synthesis routes of fullerenes and nanodiamonds, and their role in nanomedicine as drug delivery systems.

Keywords: Fullerene, Nanodiamond, Synthesis, Properties, Drug delivery

1. Introduction

Since ancient times, materials are the most interesting thing for human. In fact, human deals with different materials to meet life's requirements. Invention of new technologies requires new materials, materials with new physical, mechanical and chemical properties. Therefore, material scientists and engineers made their efforts to produce these required materials by changing their composition, structure via synthesis process [1].

In general, materials development is continuous from ancient time. So, we can notice that each period of times known by specific material's name such as: Stone Age, Bronze Age, Gold Age and Iron Age. So, due to the widespread use of nanomaterials in different industrial as well as biological fields, our current age could know by Nanotechnology Age [1].

Nanocrystalline material is one of the most interesting material with a grain size in the realm of nanometers (1–100) nm. It is not new, it is produced and used in different fields for hundreds of years. For example silver and gold nanoparticle was found in ancient paintings of glasses which gave it the ruby red color, also carbon black which is a nanostructured material was discovered in 1900 s and used in tyres technology. These materials are bigger than single atom and smaller than a microcrystalline grain. Due to their unique structure, it exhibits unique and unprecedented mechanical, physical and chemical properties that differ from those of single atom or microcrystalline grain, these materials can be synthesized from ceramics, metals, polymers and composites [2].

At nanoscales, materials undergoing to changing in their structure, physical and chemical behaviors, these changes were driven from the several effects such as; increasing their surface area, increasing atoms concentration at grain boundaries, increasing the area of grain boundaries and decreasing the percentage of dislocations within material's structure [1–3].

On the other hand, nanotechnology was presented for the first time in 1959 at the scientific annual meeting of the American Physical Society by Nobel laureate Richard P. Feynman in the famous and classic lecture that titled "There is plenty of room at the bottom". Nanotechnology can be imagined as umbrella for all different fields and areas dealing with nanomaterials and nano objects. Furthermore, it is worth mentioning that there are several observations related to the behavior of natural nanomaterials became as inspiring and promising ideas for several applications. For example: nanofluids which has been used in different applications was inspired from blood and milk (natural nanofliud) with high dispersion stability for long periods of time [1–5].

Although nanomaterials had been known and used for many years ago, but this science gained its greatest importance after the discoveries of fullerene molecules in 1985 by Kroto's and Smalley's team [3, 6], carbon nanotubes CNTs in 1991 by Saumio Lijima [3, 7] and graphene in 2004 by Andre Geim and Kostya Novoselov [3, 8]. This fact is associated to the nanochemistry of these carbon allotropes, these nanomaterials show stability did not found in any other nanomaterials due to their high activation energy barriers. Hence they attracted a great attention of researchers to employing them in a wide variety of applications even in their non-functionalized state [4].

Nanomedicine-based drug delivery system has a great influence on the targeted therapy field. Via this sector of nanotechnology, the therapeutic drugs can be incorporated into a variety of bio- nanocarriers, hence, their pharmacological behavior were enhanced and leading to high treating efficiency. Nowadays, the most attracted materials in this field is the carbon- based nanomaterials [9, 10].

2. Carbon atom

Carbon, denoted by letter C, is the most interested element in nature due to its abundant in universe. It is 15th most common element on earth's crust [11]. In nature, carbon's ores can be exist in different forms, as carbonates [11], wood charcoal [11], briquette and others. In human body, it is the second abundant element by mass after oxygen. Also, carbon atoms exist in all organic materials. So, it is regarded as the basic building unit for all life [11].

In periodic table, carbon atom exist in group 14, therefore it is a tetravalent. Carbon has fifteen different isotopes, two of them are stable known as ^{12}C and ^{13}C that comprise 98.89% and 1.11% of the carbon on earth respectively. Other isotopes known as a radio-active isotopes, among them ^{14}C is the more stable one [12, 13].

In general, the term of electronic configuration refers to the electronic structure of an atom, with its levels, sub-levels and electrons number that occupying its orbitals. The electronic configuration of carbon atom is: $1s^2\, 2s^2\, 2p^2$. Excited configuration is the basis of the hybrid orbital. $1s^2\, 2s^1\, 2p_x^1 2p_y^1\, 2p_z^1$, hence with four unpaired electrons, four L- shell can be hybridize then and forming sp^3 orbital which directed

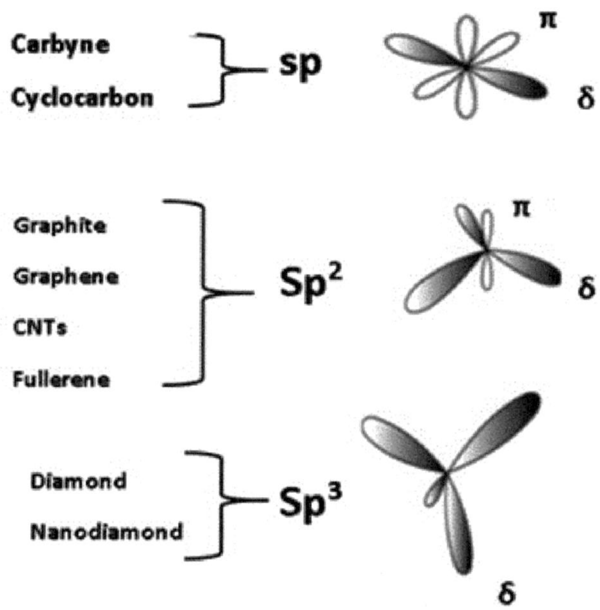

Figure 1.
Representation of carbon structures hybridization states.

in the tetrahedral direction at angles of 109.5°. This electronic structure found in diamond. When 2 s and two of 2p orbitals hybridize together, three of the sp^2 orbital was formed and directed in the trigonal direction at angles of 120° from each in a plane. This electronic structure found in graphite. While in sp. hybridization type, there is only one of 2p orbital is used which led to diagonal bonds with 180° apart. This electronic structure found in carbyne. Also it is good to mention that the unused 2p orbital in each sp^2 and sp. hybridization types forming π bonds [14, 15]. See **Figure 1**.

Carbon atoms have the ability to form long chain with its own atoms via feature called "catenation". There are three basic covalent bonds between carbon- carbon atoms, single bond (sigma bond) is the most common bond between carbon–carbon atoms, and it is consisting of one electron from each carbon atom and is thus a two-electron bond such as the bond in ethane. A double bonds is another type that is exist between two carbon atoms, generating the alkene group, also called olefins. These hydrocarbons are exclusively composed of carbon and hydrogen. The third type of bonding between tow carbon atoms is the triple-bond which is existing in compounds called alkynes. Due to the flexible chemistry of carbon atom and the strength of the covalent bond, carbon has the ability to create endless chains, sheets and three dimensional structures with different properties [14, 16].

3. Carbon allotropes

The most prominent allotropes of carbon are graphite, diamond and lonsdaleite. Each one has its specific properties which very varying from that of another one [16].

Graphite was named by German mineralogist and chemist A.G. Wenner in 1789 [11]. It is a naturally occurring material found in a metamorphic rocks. Its structure made by holding many layers via weak forces. Each layer consisting carbon atoms arranged themselves in a honeycomb lattice [16]. Graphite is characterized by its dark opaque appearance with soft texture and well electrical conduction property. Within its structure, three electrons of each carbon atom shearing with another neighbor by single bonds while the fourth electron become delocalized overall the whole sheet of atoms in each layer. The delocalized electrons are not fixed to a specific carbon atom and can be move freely on the sheet's surface leading to induce a temporary dipoles which induced an opposite dipoles above and below each sheet that leads to hold the overall sheets of graphite structure [17].

Diamond was derived its name from the Greek word "adams". Diamond is the cubic high pressure phase of elemental carbon [18]. It is occurred naturally and it's the hardest known material [18], with a transparent appearance [11] or can be exist in different colors depending on the presence of specific elements or some structural defects with a very low electrical conduction [18]. In diamond, carbon atoms forming 3-D lattice structure in which each carbon atom shears its four electrons with another carbon atom by four single bonds [11].

Lonsdaleite, the hexagonal diamond [11], it is named in honor of Kathleen Lonsdale. It is found naturally in a graphite containing meteorites that struck earth. It is a transparent material with hardness much higher than that of diamond [11, 16].

Amorphous carbon [19], is an interested allotrope, in which carbon atoms can be exist with a wide variety of hybridization states. The most familiar form of amorphous carbon is diamond-like carbon (DLC) due to its structure which is amorphous with many fraction of sp^3 bonding so it exhibit some important characteristics of diamond without the necessary for complicated synthesis techniques [20]. It is characterize by its high hardness, low friction property [20] and chemical inertness [20].

Furthermore, there are several allotropes of carbon atom exist in nano- scale featured by their unique structures that led to a great explosion in materials science and technology. Here we will give a summarize description about the structure of some of these materials [21].

Graphene, attractive nano- scale allotrope, is an atomic thick sheet made up of carbon atoms arranged themselves in a honeycomb lattice. So, it is a 2-D allotropic form within carbon family. It is represents the structure of other allotropes of carbon due to its structure flexibility. By rolling its structure, carbon nanotubes is obtained and by warping up a fullerene allotrope is obtained while the stacking of several sheets leads to produce graphite structure. So it is called a mother of all other carbon allotropes, see **Figure 2**. Due to its attractive properties it is candidates for many critical applications in industrial [4] and health care fields [22].

Carbon nanotubes is a cylindrical structure nano allotrope of carbon with a diameter of several nanometers and micro- scaled length, consisting of rolled graphene sheets. Its structure can be vary by its diameter or, its length or its layers number.

There are two main types of CNTs, single walled SWCNTs and multi walled MWCNTs. Several studies had been clarified that SWCNTs have about (1–3) nm diameter and few micrometers in their length, while MWCNTs have about (5–40) nm diameter with length around 10 micrometers. This type of carbon allotropes have unprecedented aspect ratio property due to their unique structure together with high thermal and electrical conductivity compared to other conductive materials [4, 23].

Fullerene and nanodiamond another allotropes of carbon atom, have unique and novel properties due to their unique nanostructure. These two carbon nanomaterials will be discussed later.

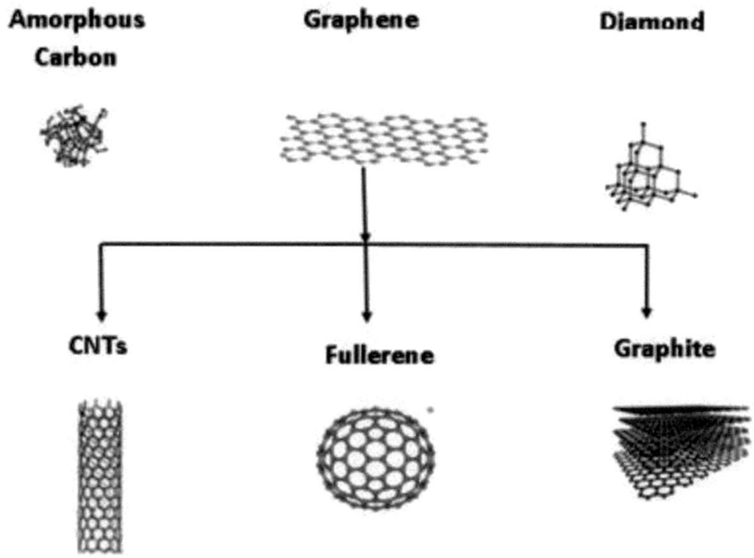

Figure 2.
Representation of carbon allotropes.

4. Hybridization of carbon atoms

The recent materials discoveries at nano scale open new horizons in all science and engineering fields. One of the most important point of view reported in several literatures is depending on classify carbon materials according to their hybridization characteristics [24, 25].

The electron hybridization determines the ability of carbon atoms to arrange themselves in a wide variety of structures with linear, planar and tetrahedral symmetry forming different allotropes, such as carbyne/carbyte, graphene/graphite and diamond. So, depending on hybridization type carbon structures can be classified from sub-molecular to macroscopic scales into three general families: carbyne- general family (sp^1 family), this family includes carbon atoms with sp^1 hybridization state from its nanostructured like carbyne or cyclo-carbon to its macrostructural crystalline form [26]. The second class called graphene general family (sp^2 family), this family includes carbon atoms with sp^2 hybridization state in their nano as well as macro-crystal structures with hexagonal and rhombohedral structures like graphene and graphite. The third class called diamond general family (sp^3 family), the carbon within this family existing sp^3 hybridization state like diamond structure in its nano and macro scales [27].

For more complexation, carbon atoms can be exist with other hybridization states called intermediate hybridizations with different degrees. The degree of hybridization in this case caused by the curvature of the sp^2 hybridized structure, leading to produce strained C–C bonds. Usually carbon atoms with pure sp^2 hybridization state shows an ideal flat structure. While in the case of atoms with curved structure, their hybridization degree should be >2 as in the case of fullerene carbon [14, 27].

On the other side, usually carbon atoms in sp^2 hybridization state arrange themselves in hexagonal rings as well as in a various other polygonal rings. The non-hexagonal rings can leads to curving the flat sheet or keeping its flatness if the polygon arrangement fulfils certain symmetry rules [14].

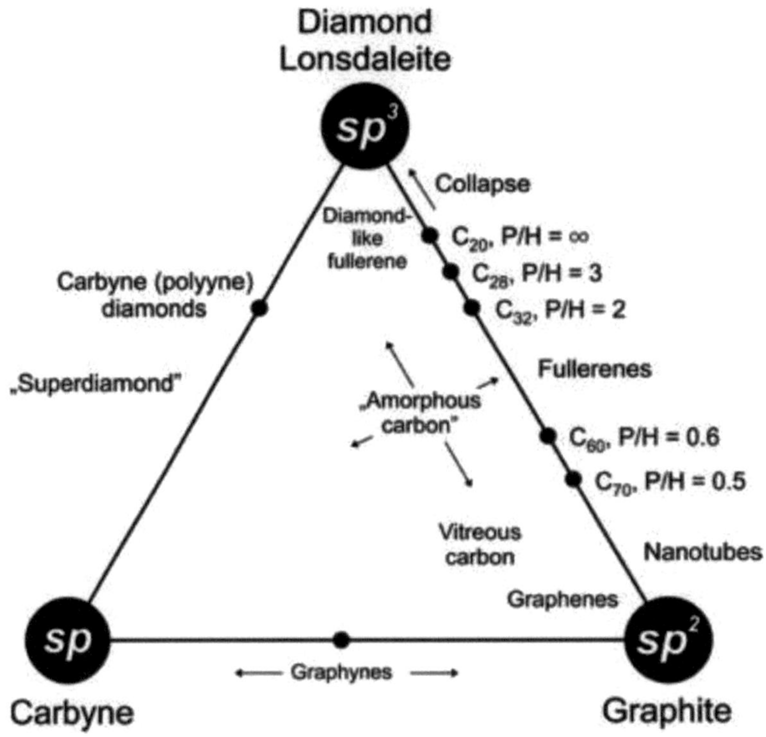

Figure 3.
The ternary carbon allotropy phase diagram based on hybridization type (reuse with permission Elsevier [14]).

The pentagon rings induce a positive curvature while heptagon or octagon rings induce a negative curvature. Therefore, carbon materials with sp^2 hybridization state can be exist into three types: positive curved fullerene-type, which includes carbon atoms with hexagonal and pentagonal rings and negative curved schwarzite-type, this type includes carbon atoms with hexagonal and either heptagonal or octagonal rings and the last one called zero-curvature graphene-type which includes carbon atoms with hexagonal rings only [14]. This procedure of classification carbon based- nanomaterials had been presented by a triangular carbon allotropes phase diagram, see **Figure 3**. In which, hypothetical carbon allotropes located at the corner of the diagram, and allotropes with intermediate hybridization states located at the edge of the diagram while carbon existing mixed hybridization states located inside the triangular diagram [14].

5. Fullerene and nanodiamond

The recent discoveries of the nano-sized carbon materials leads to expand the list of carbon allotropes. In association with the ability of carbon atoms to form wide range of structures, carbon based- nanomaterials become widespread in the fields of nanoscience and nanotechnology [28].

Generally, Carbon-based nanomaterials possess effective physicochemical properties make them as a powerful tool in medicine. For example, graphene possess many promising properties due to its high surface area and high functionalization ability that make it suitable for drug delivery treatment, along with its high mechanical properties, graphene had been recommended for tissue engineering field. Carbon nano tubes CNTs had been suggested for different in vivo applications due to its strong optical absorption in the specific wave length and as an active tool for bi0 imaging and drug delivery applications. Recently, fullerene and nanodiamond had been investigated and received much attention to use as a drug delivery carriers [29].

In spite of the promising benefits of using carbon based- nanomaterial as medical tools for the treating of difficult to treat diseases, several challenges are involved within this technology such as, their toxicity, diffusion and distribution abilities throughout the body which may leading to unpredictable effects. So, in-depth and carefully studies around their nature and behavior in human body regarded the most important factors in this field. One of the great promises of nanotechnology in medicine is the local or targeted delivery of drugs. Efficient targeting would allow for a reduced systemic dosage meaning also a reduced toxicity while resulting in relatively higher or more efficient dosage at the desired target site [29].

6. Fullerene

6.1 History and structure

Several ideas, suggestions and observations in addition to physical and chemical experiments of clustered molecules were led to the way for discovery of C_{60} in 1985.

In 1966, David Jones discussed the possibility of creating balloons made up from carbon atoms. Then in 1970, this idea was progressed by Eiji Osawa when he revealed the possibility of preparing molecule made up of 60 carbon atoms knowm as C_{60} molecule in a condensed icosahedron structure [30]. After that in 1971, Eiji Osawa and Zensho Yoshida enumerated the possible aromatic properties of the structured C_{60} molecule. Later, Bochva and co-worker studied the electronic structure of this molecule. Then in 1980, Davidson characterize the closed- hollow structure of this molecule using different theoretical techniques. Subsequently, in 1985, Kroto and Smalley and their team obtained carbon cluster through scientific experiment to study the suitable conditions at which carbon atoms nucleates in the atmosphere of the red gait star. The mass spectrometer analysis of the obtained clusters indicates that most of them had 60 carbon atoms and some of them had 70 ones [31].

This carbon allotrope become the heart of nanotechnology and attracted significant attention of scientists. For that, in 1996, Kroto, Curl and Smalley were rewarded by Nobel Prize in chemistry for their discovery of fullerene [31].

Fullerene derived its name in the honor of the geodesic domes designer the artchitect Buckminster Fuller. Fullerene family usually represented by a formula of C_n, where n refers to the existing carbon atoms in the cage structure which can be up to several hundred atoms, the number of the carbon atoms existing within fullerene structure has a great influence on its structural geometry and its properties. C_{60} is the most dominant molecules within fullerene family [31].

The structure of C_{60} has truncated icosahedrons made up of 20 hexagonal rings located at the center of the icosahedral faces and a 12 pentagons located around the apexes. It is the most symmetric molecule. It has 2.fold, 3.fold and 5.fold rotational symmetry. The first one is through the edge center of 2- hexagons, the second one is from the center of 2-hexagons which facing each other, while the last one is through two pentagons centers which facing each other [30, 31].

Furthermore, fullerene molecules can be exist as a spherical, ellipsoid, tubular shapes consisting hexagonal, pentagonal and sometimes heptagonal rings. C_{60}

belongs to the spherical fullerene class which looks like a soccer ball, while C_{70} belongs to the ellipsoidal class which looks like rugby ball. In addition, several efforts have been reported to produce fullerene with high yields, in 1990, a method was discovered for producing macroscopic amounts of this distinctive material and this breakthrough allowed scientists to understand its chemistry and explore its properties [31].

Generally, fullerene can be classified into classical fullerene and non-classical fullerene. The first one containing 12 pentagons and any number of hexagons, while the second fullerene class can have heptagons, octagons, and an additional number of pentagons or squares [30, 31]. Due to the unique structure and properties of C_{60}, scientists showed high interesting in synthesis both larger and smaller fullerenes. Therefore, the family of fullerenes has been expanded involving fullerene molecules with a wide range of carbon atoms number. Larger fullerenes that have an icosahedral- symmetry also can be constructed. While, the carbon cages structure smaller than C_{60} consist of adjacent pentagons. These smallest fullerenes are predicted to have unusual physical as well as mechanical properties due to the high curvature of their molecular surface. The smallest fullerene molecules is a dodecahedron consisting of 20 carbon atoms with only pentagon rings. The fundamental understanding of the size dependence of the closed carbon cage structures is important for tailoring these systems for possible nanotechnology applications [31].

Due to the electronegative nature of fullerene, fullerene can form different compounds with different structures. One of the most important fullerene species is derived from the cage- like structure in which there is an ability of trapping metal atom inside the cage and forming specific endohedral fullerene known as metallofullerene. Exohedral fullerene is another type of fullerene with enhanced solubility property, obtained due to the chemical reaction with chemical groups. On the other side, when one or more carbon atoms within cage structure are substituted by specific hetero atom a hetrofullerene is produced [31].

6.2 Synthesis routes

Synthesis of graphenic materials like fullerene have been studied and reported in many literatures. Different techniques have been adopted for fullerene synthesis, such as arc discharge technique [32], vapor deposition of carbon atoms technique [33] and laser technique using graphite [34]. The cage- like structure of fullerene was identified for the first time by Kroto and Smalley in 1985, their experiment depending on applying an intense pulsed laser on a rotating graphitic disk to vaporize carbon atoms in the presence of helium atmosphere, then the condensed material had been collected, some of these routes are shown in **Figure 4** [35].

In 1990, another method was used to prepare fullerene reported by Ajie and co-workers, this method depending on the principle of the resistive heating of carbonic rods in a partial helium atmosphere which leads to evaporating carbonic atoms and then condensing it into fullerene structure [36].

Another method was discussed in U.S Patent in 1991, using electric arc technique, through this process an electric arc is generated between graphitic rods in inert atmosphere leading to produce soot- like product in which fullerene molecules extracted from the soot using suitable solvents [37].

The previous methods were associated with producing fullerene with low yields and there is no temperature controlling zone which is required for graphitization step. To overcome these two limitations, in 1994, Smalley discussed a laser vaporization technique to prepare fullerene from graphite materials using a focused laser beam. This technique involving evaporating carbon atoms and retained it in a

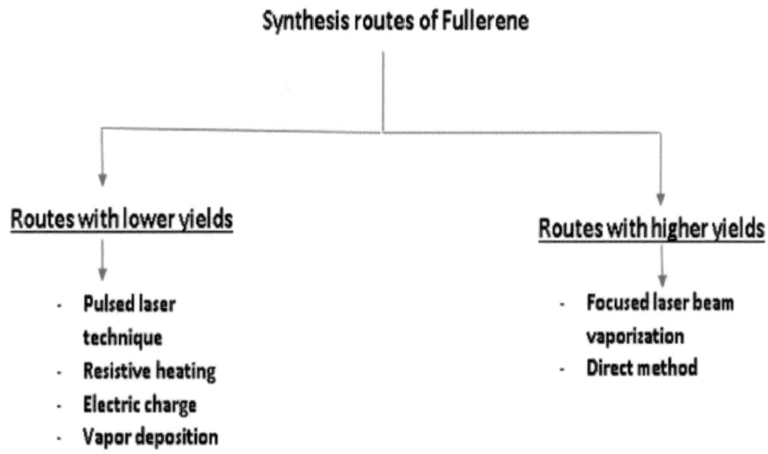

Figure 4.
Some synthesis routes of Fullerene.

temperature- controlled- zone for sufficient time in order to complete the growth and annealing process of the produced structure [38].

Furthermore, another method was reported by Boorm and co-workers in 2001 called the direct method. This method depending on using polycyclic aromatic hydrocarbons with fullerene- like- framework. Through this method fullerene was synthesized directly due to rolling- up of the hydrocarbons structure into fullerene structure via laser irradiation process under flash vacuum pyrolysis conditions [30]. Whereas, in 1993, fullerene has been synthesized in high yields via ablation technique of graphite rod with solar irradiation using solar- furnace. Through this technique graphite was vaporized under the action of direct exposure to high- flux- solar- irradiation. Then, the carbonic vapor was entrained by helium flow and cooled into dark- water- zone to form fullerene structure [39].

Since the discovery of fullerene as an important carbonic allotrope, tremendous development have been made in nanoscience and nanotechnology in order to fill the urgent need for this materials in a wide variety of applications. But, the formation mechanism of the cage- like structure still as a mystery and not well known [36].

One attempt to explain fullerene formation mechanism was reported by Paul and co-workers in 2012. Through their experiment, they depending on the principle of the bottom- up technique to explain the growth mechanism of fullerene by integration with carbon atoms and C_2 using laser irradiation technique. A carbonic target made up from graphite or ^{13}C amorphous carbon with fullerene content exposed to a single laser strike that leads to vaporizing carbonic target into atomic carbon and C_2while fullerene molecules desorbed into carbonic vapor zone under the action of helium flow. Then an interaction takes place between C_{60} fullerene (98.9% ^{12}C and 1.1% ^{13}C) and enriched ^{13}C carbonic vapor which involves the exchanging of the original ^{12}C atom of fullerene structure with ^{13}C atom of carbonic vapor [40]. Each ingesting of ^{13}C atom into fullerene structure occurs parallel with ejecting ^{12}C atom from its structure and this interaction will spurring bond rearrangement with fullerene structure. Then, the produced species leaves clustering zone and undergo to a supersonic expansion step. Hence, different fullerene isomer such as C_{70} can be formed and the ejected ^{12}C atoms may be ingested or exchanged with another fullerene structure [36].

Furthermore, the growth mechanism of fullerene via different techniques that rely on the interaction between carbon atom and C_2 affected by several parameters such as the density of carbonic vaper, exposure time, the flow rate of inert gas, therefore, there is a possibility of controlling the growth mechanism by these parameters [36].

7. Nanodiamond ND

7.1 History

Recently, nanodiamond ND with its novel structure opened a new path in the creating and developing materials field. Nanodiamond is a new member of carbon- base- nanomaterials consisting nano tetrahedral network [37]. Naturally, nanoparticles of diamond have been found in meteorites as well as they have been found as inclusions in the old crust fragment of the earth [38].

The history of nanodiamond discovery starts since the second half of the last century via several successful and unsuccessful attempts in the synthesis and analysis of nanodiamond ND or ultra-dispersion diamond UDD begins in 1956 by Yu. Ryabinin using dynamic synthesis approach [39]. In 1961, diamond have been detected in a preserved shock compressed graphite in a plane ampoule by P.J.De Carli and co-workers. Then, in 1962, diamond was produced via shock compression using carbon black and graphite as starting material in cylindrical and spherical storage ampoule with up to 2% yields. Later, diamond was synthesized in explosion chamber using graphite, through this process graphite was placed directly into cylindrical charge containing of a troty- hexogen mixture and the charge was enveloped in a water jacket in order to inhibit graphitization process.

During 1963–1965, the effect of explosion conditions on the produced UDD was studied and indicate that the DP cooling as a result of conversion of the potential energy of diamond particles into kinetic energy of the envelope surrounding the charge plays a decisive role in the UDD synthesis [29]. Then, in 1976, Dupon co. used a cylindrical ampoules to produced diamond micropowder in commercial production rates by compressing a mixture made up from graphite and copper with charge. After that several scientific experiment were carried out to produce UDD with high yields using large mass charges. Another attempts were carried out in 1994 by V.V.Danilenko and co-workers to sintering UDD under static conditions [39].

Furthermore, a wide variety of techniques have been reported and employed to prepare micro and nano diamond particles. But on the other hand, the main drawback of most of these techniques are their requirement for high temperature and high pressure conditions, also, it was found that the produced material is a mixture made up from diamond and nan-diamond phases and some procedures leads to precipitate a amorphous carbon films at the grain boundaries of the produced nanodiamond [40]. After that, several techniques were suggested to produce nanodiamond at lower temperature and pressure conditions [41].

In this area, many researchers compete to synthesis ND with higher yields using advanced techniques in order to use it for a wide range of applications such as in drug delivery, biotechnology and tissue engineering fields [42].

7.2 Structure and properties

ND have unique properties, for this reason it is attracted the desire and interest of scientists and researchers in the physics and chemistry of nano- materials. It is believed that the structure of ND consisting of single or more diamond crystal surrounding by a shell containing graphenic carbons sp2C, amorphous diamond sp3C in addition to surface- state- carbons. The latter made up from chains of trans- polyacetylene TPA and graphene/fullerene fragments. In addition, different functional groups have been found, see **Figure 5** [43].

However, some of the existing sp2carbon and amorphous diamond can be discarded by specific techniques such as thermal oxidation technique [44], while some graphene/fullerene fragments of the shell are an intrinsic components formed through the re-arrangement of the diamond surface. So, it is believed that the collected information about the structure is necessary in order to understand the nature of interaction between ND particles on one side and with other compounds on another side [43]. It is worth mentioning that the shell's constituents will have a great influence on the properties of ND particles with smaller sizes [45].

In the last decades, it was supposed that the shape of ND particles were quasi-spherical, but the modern microscope confirmed they are a polyhedral with distinct faceting shape and about half of the presented carbon atoms in ND are located at their surfaces. Hence, ND particles have bonding ability with different functional groups that effecting on its stability [46]. ND particles were required for a wide variety of applications, in lubricant industry, composites, medical therapy and others, this is due to the nature of their surface chemistry which depending on the chemical history of the material and the synthetic process [47]. For example, oxygen- rich- functional groups like hydroxyl, lactone and carboxyl have been found on the ND particle's surface produced via detonation technique [48]. Several efforts were reported about the surface modification process by functionalization with different groups which regarded as an effective strategy for reducing the size of ND aggregations. For example, functionalization with long chains of alkyl leads to reducing their aggregation size and enhancing their dispersion ability in organic solvents. Similar effects have been achieved via functionalization with boran. Furthermore, functionalization with Lysine molecules showed better water dispersion ability comparing carboxylic- ND [48].

Another type of ND is called hydrogenated ND, in which the surface of nanoparticles were wholly hydrogenated, hydrogenation process involves of linking a hydrogen atom with carbonic specie, then hydrogen atom will take its active role in etching of ND particles such as graphenic/fullerene carbon sp2 C or amorphous carbon, discarding oxygen-rich-groups as well as forming C-H bonds at ND surface [47, 49]. In fact all these benefits make this type more attracted for the most critical applications.

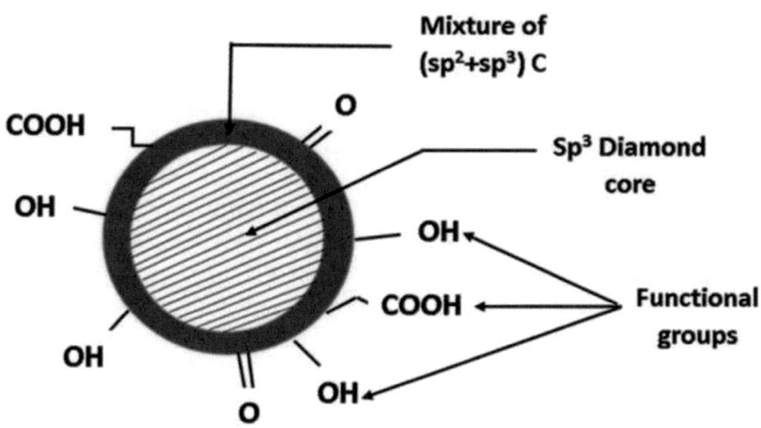

Figure 5.
Representation of ND structure.

7.3 Synthesis routes

Graphite is the most stable carbon allotrope at ambient conditions of temperature and pressure. While, diamond formation requires more severe conditions. For example, the conditions for naturally formed diamond are (>1000°C) temperature and (4.5–6) Gpa pressure. After diamond formation, the reverse transition to graphite structure will not occur due to the high- energy- barrier for phase transition. Although, graphite is the favored allotrope from thermodynamic point of view, but about 0.4 eV energy barrier must be overcome to transfrom sp2 C structure to sp3 C structure and this fact makes diamond as a metastable allotrope. But the transition kinetic to graphitic structure is not allowed [50].

Nowadays, a wide variety synthetic techniques for ND are available. In the following a brief description of the main techniques.

1. Detonation technique.

2. Chemical Vapor technique.

3. Laser technique.

7.3.1 Detonation technique

In detonation technique an explosives with a negative- oxygen- balance and a source for carbon atoms (graphite or molecules driven from the used explosive materials) were placed inside the detonation chamber which is a closed metallic chamber. The driving force for diamond formation obtained from the explosion energy. During detonation, carbon atoms released and then condensed and transform into nanoclusters of crystals. In association with the generated high pressure and temperature, a crystallization of nanoclusters will occur and ND particles will form and grow into aggregations with size about (4–5) nm. The used coolant agent can exist as gas (dry detonation) or as water (wet detonation) [38, 39, 49].

The final produced soot-like material is usually consisting of diamond core with sp3 C surrounded by sp2C. The main advantages of this technique is its ability to produce ND with wide range of particle size, structure and surface- functional- groups these features makes this technique useful for a wide variety of applications as polymer filler for nanocomposites, polishing and coating purposes and others. While on the other side, the main associated disadvantage is the contamination of the produced ND with fragments from chamber wall [49]. So in order to eliminate the unwanted sp2C, to discard metal contamination and to breakup ND aggregations, a post- treatment step will be an essential to produce pure, de- aggregated ND with sp3C [39, 49].

Formation mechanism of detonation ND has been proposed by Danilenko. He suggested that the required temperature to form liquid carbon from its nano scale is lower than that from its bulk scale. So, he is suggested that the liquid carbon region is shifted to low- temperature region while the stability region of ND is shifted to high- pressure. This situation leads to a homogeneous nucleation of ND in carbon supersaturated vapor region followed by crystallization of the produced carbonic liquid [50].

7.3.2 Chemical vapor deposition technique

Till this day, CVD technique is widely used in the preparation of ND as powder or thin film with a wide range of sizes. The required carbon atoms for ND formation is derived from the decomposition of gaseous phases (usually methane in hydrogen excess) and a carbon-rich species. During decomposition process, the released carbon atoms were deposited on a silicon substrate covered with a detonation nanodiamond DND acting as a seeding sites for nucleation of ND [51].

For further illustration, several sources of energy can be used to activate the gas phase such as hot filament [52], plasma and flames [53]. Due to decomposition of gas phase, radicals will forms and then each two adjacent carbon atoms located at the surface of the used diamond-coat are left with the dangling bonds after hydrogen abstraction by H• radical. After that these bonds will be full with CH3 radicals then the adjacent carbon atoms will bond together and locked within diamond lattice [54, 55].

The main advantage of this technique is its ability to produce ND particle or ND thin film with a wide range of size (10–200) nm, in addition, to the possibility of controlling on the structure and morphology characteristics of ND. Also there is possibility to produce ND doped with different species that can inserted within diamond structure through growth process. In fact these possibilities enabled this technique to produce ND with modified electrical and optical properties required for many applications [54, 56].

7.3.3 Laser technique

This technique is one of the most attractive used technique for synthesis ND in liquid [38]. This technique including directing an intense laser beam on the graphite target immersed in a liquid medium, usually water. The directed laser beam with high energy induces target surface melting and turning it into superheated liquid. Due to the highly increased temperatures, a phase explosion will takes place and nanodroplets will be forms [57]. At these conditions, the emission of plasma plume with ablation will occur and an extremely high pressure and temperature conditions are created. Then through cooling of the ablation plume with the liquid medium a rapid quenching will takes place. In fact this situation of rapid and sudden decreasing in temperature creates the appropriate conditions for carbonic nanodroplets formation within few nanoseconds [49–59].

This technique presents good benefits such as the ability to produce ND with high purity, while the main drawback is the high cost with low production rates [39]. On the other hand, several attempts have been reported to overcome such undesired features one of these attempts is called (Light Hydro- Dynamic Effect) LHDE. This technique used laser beam with higher power cross a fluid with a specific refractive index. The direction of laser beam produces white light flash and generates acoustic waves which leads to form high- power- hydro- shock [39].

This technique produced ND with high yields in association with good controlling on its size and surface functional groups. Moreover, it is found that the produced material possess outstanding thermal property make it suitable for nanocomposites applications require heat dissipation property [58].

7.4 Nanodiamonds at ambient conditions

Nano-sized diamond have been synthesized via different techniques as previously discussed. In spite of the expected stability of the produced material, high-temperature and high- pressure are the main requirements for these techniques. In addition, the produced material is usually consisting sp3 and sp2 carbon and some of these techniques leads to produce ND with contaminates which requires additional purification steps and hence the overall cost will be increase. Therefore, several studies have been reported to prepare ND at ambient conditions [60, 61].

Proceeding from the chemical principles, the chemical reaction depending on the energy- materials interaction degree and require specific energy to proceed. Some of these studies suggested the possibility of using ultrasonic irradiation technique as a source for the required energy [60].

So, ultrasonic irradiation technique has been candidate to prepare ND. This technique improves the chemical reaction in solution via production of hole-constituent micro-bubbles [60]. During this process, a liquid media is irradiated with ultrasound and bubbles will be creates and oscillates under the action of the alternating compressive and expansive acoustic waves. Then, these bubbles will grow to a critical size and collapse leading to release an intense- localize energy about (5000 K and 1000 bar) within a very short period of time which is enough for synthesis of nanodiamond particles [61, 62].

The features of this technique were taken by researchers to produce micro and nano diamonds. In 2008, Khachatry and co-workers used graphite organic liquid suspension to synthesis microcrystalline diamond with degree of purity, cubic-structure and size range about (6–9) μm via ultrasound- cavitation process [63].

In 2019, researchers used a suitable method to synthesis nanocrystalline diamond via ultrasound waves. Through their experiment, graphene oxide was synthesized by modified Hummer's method then the prepared GO dispersion is activated by ultrasound waves which impacts its morphology and chemistry leads to produce graphene sheets. On the other hand, the generated shear forces leads to convert some of the produced graphene sheets into graphenic nanoscrolls with Mn2O7 had been inserted within its cavity. After decomposition of the unstable Mn2O7 a localized damage at the nanoscroll structure takes place which then undergoing to self-healing and as a result, ND seeds was formed and warped into nanoclusters of diamond, see **Figures 6** and 7 [62].

Other studies have been shown that the nucleation of ND is preferred inside the structure of carbon nanotubes under the action of surface tension property as a result of the carbon nanostructured curvature [64]. In association of founding some carbonates such as Li2CO3, Na2CO3 and K2CO3 as inclusions in natural diamond. Kamali and co-workers made their study in 2015 to produce nanocrystalline diamond from lithium- carbonate containing nanostructured carbon by simple heat treatment and at atmospheric pressure. In this study, CNTs produced by electro-chemical process in lithium chlorite melt using graphite material. Through the experiment, lithium ions discharge on the cathode and inserted between the layers of graphene of the graphite structure under the influence of the cathodic which led to initiated enough stress to pulling graphene sheets from graphite structure into the melt and then these sheets will rolling up into CNTs. And after simple oxidation of the produced CNTs at temperatures range (420–550) °C and at atmospheric pressure, ND was formed, see **Figure 8** [64].

Another methodology was reported by Maia and co-workers in 2015. Their strategy depending on using dynamic compression as a tool for carbon structured transformation process under the action of accumulation of ultra- short laser pulses assisted by the formation of onion-like carbon structure as intermediate phase. Basically, the accumulation of free-electrons with high densities at the grain boundaries of graphite absorbed the applied energy which leads to creation a super ex cited region and then an ablation takes place followed by the propagation of non- thermal shockwave. After that, heating and thermal equilibrium takes place. In fact, these sequences of effects happened at each strike of laser which leading to destroy the lattice in continuous manner and causes more carbonic order to form. The formation of latter structure considered as intermediate phase with lower energy barrier to allotrope transition while the degree of crystallinity increases at each strike of shockwave [65, 66].

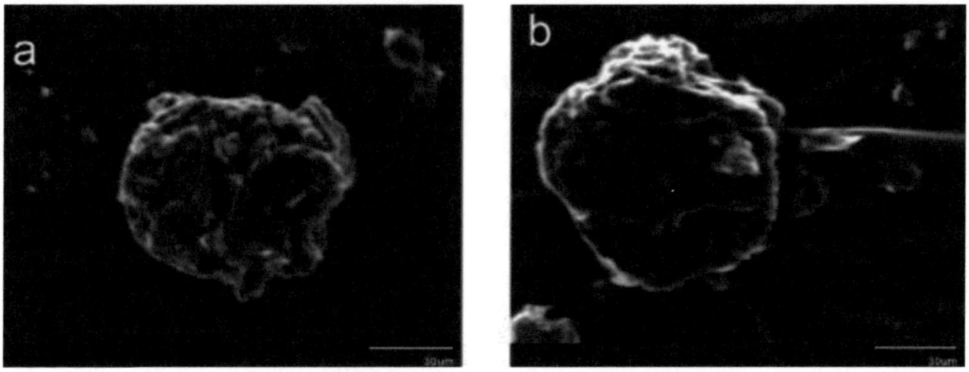

Figure 6.
SEM images of nanodiamond at its cauliflower stage.

Figure 7.
(a-b) SEM image of Nanodiamond ND.

Figure 8.
SEM images of the micro and nano sized diamond particles after heating CNTs to 550°C in air (reuse with permission Royal Society of Chemistry [64]).

8. Drug delivery system

Nanotechnology is the nano-sized materials science, involves materials manipulation at atomic and molecular scale in order to produce materials, systems and devices with unprecedented features. Recently, nanotechnology become the most promising technology in a wide variety of fields, one of these fields is medicine field. In fact, the employment of nanotechnology in medicine field is called nanomedicine which regarded as a tool enabled doctors to reach the human body at the molecular and cellular levels and treat the damaged tissues [67].

It is believed that this technology will have a great impact on the health care field via its effect on the sickness diagnosing and treating [67, 68]. There are several features of using nanotechnology in medical field, for example, when it is used for drug delivery, this technology will protect drug from degradation within human body before reaching the target also, it is improving drugs absorption into the diseased cells and at the same time it give assurance that there is no any interactions between drugs and the healthy cells [67, 68].

Therefore, nanomedicine is one division of nanotechnology and nanoscience has the ability of treating the diseased and damaged cells or organs within human body at the cellular and molecular levels via nano-devices and nano-structured materials [68]. There are three main sectors within nanomedicine: Nano- diagnosis, which involving detection and analysis of the diseased cell using different devices such as imaging devices. Nano- therapy, this sector involving the direct transfer or delivery of drugs to the diseased cells with the least possible of side effects. And the last one is renovated medicine, which involving fixation and replacement the deteriorated parts within human body using different nano-devices and nanomaterials [67].

In the following we will discuss the role of carbon based- nanomaterials (fullerene and nanodiamond) as a drug delivery systems.

9. Significance of carbon based- nanomaterials in drug delivery

Generally, the suitable choice of nano- drug delivery systems aids to overcome many health issues usually associated with using traditional treatment strategies. For example, in the case of cancer chemotherapy, the traditional strategies leads to several undesired side effects such as suppression of bone marrow, hair loss, gastric and renal damage and other toxicity effects [67].

On the other side, there are many features and reasons related to the nature and structure of nanomaterials make it as an attractive subject for intense bio-studies from one side and as an attractive materials for drug delivery systems from a other side [69, 70]. The most significant features are: their quantum property, their sizes which determine their in vivo and in vitro behavior, their structure and aggregation ability as well as their surface atoms or molecules. In fact, all these features determine their ability for binding, carrying and adsorbing other compounds or in other words these features determine their pharmacology behavior [70–72].

Furthermore, the main characteristics that nanomaterials should have to be use in drug delivery systems are high- solubility, bio- compatibility, bio- availability, bio- distribution and targeting ability, drug incorporation and release ability, their shelf- time, anti- clotting property and bio- degradability [71].

9.1 Fullerene-drug delivery system

Some of the most interesting characteristics of fullerene are their size, electronic configuration, hollow and cage structure, their inertness and surface modification ability that offer the utilization of using fullerene in biological and medical chemistry fields and open new horizons in nanomedicine [73]. The main problems facing the previous possibilities are the insoluble nature in aqueous media with high aggregation tendency [73]. But on the other side, there are several attempts have been done to overcome these problems. One of these attempts involving encapsulation in specific carriers such as calixarenes, micelles and liposomes. Other attempts used chemical functionalization methods with carboxylic acid, polyhydroxyl and amphiphilic polymers to increase the hydrophilicity property of these structures [73]. Furthermore, fullerene shows a nontoxic behavior which tend to decreasing with increasing surface functional groups, while the circulation and biodistribution property depending on the composition of the existing derivative groups. In addition, the presence of functional groups acts as a flexible interfaces for tuning the required drug delivery and its action besides the size of fullerenes even with their derivatives are still smaller than other types of nanoparticles [73, 74].

Both exohedral fullerene which have additional atoms, ions, or clusters attached its outer spheres structure and endohedral fullerene which have additional atoms, ions, or clusters enclosed within its inner spheres structure, have been employed in nanomedicine field as drug carriers. Endohedral fullerene and its derivatives can used to deliver atoms or ions in biological systems, for example, metallofullerene can be serve as drug delivery depending on the composition and properties of the trapped metal within its structure [75]. Depending on the type of functional groups, exohedral fullerene can be exist in three main forms. The first one is called surface- derivative fullerene, the biological action of this type is driven from the inherent properties of its structure such as: their size, reactive property and photochemistry property. The candidate application for this type is as antioxidants systems due to its electronegative nature in association with its reaction ability with different radicals. So, this type work as antioxidants radicals' scavenger. Furthermore, surface- derivative fullerene can be used to generate a reactive- oxygen species by light irradiation, so this type is useful as photodynamic therapy for killing cancer cells and other undesired cells [75].

The second type of exohedral fullerene is known as covalently fullerene. In this type, the derivative surface of fullerene is directly connected to the pharmaceutical activated compounds via covalent bonds. This fullerene molecule has unique biofunctions with biological body such as drug release with selective targeting at the same time. For example, C60- paclitaxel fullerene used in lung cancer treatment [76], here the paclitaxel has been conjugated to C60 derivative via hydrolysable-ester- group linker used for slow release of drug to the diseased cells [76]. Furthermore, this type of fullerene has high tendency to attach covalently with different biological relevant like sugar, cholesterol, carbohydrates and others [76].

The third type of exohedral fullerene is called non- covalently fullerene. Here, fullerene and its derivatives tend to form a complex species via a wide variety of interactions. These interactions could be exist as pi-pi interaction, van der Waals interaction, electrostatic interaction, hydrophobic interaction and it could be exist as a hydrogen bonding. An example of using this type of fullerene in nanomedicine is the protease inhabitation of Human Immunodeficiency Virus (HIV) [77] by C60 fullerene of this type. The active catalytic sites of HIV protease enzyme is as a hydrophobic bag with a 1 nm diameter which is very close to the size of C60 molecule. Hence, this fullerene will bind with the active sites and virus inhabitation process will takes place [75].

9.2 Nanodiamond-drug delivery system

Nanodiamonds are one of the most interesting material in nanomedicine have the ability to conjugate with different drug molecules [48]. This fact is born out of their unique structure and properties. High surface flexibility, small size, high surface area and functionalization ability with different molecules besides their biocompatibility which is more than that of other carbon based- nanomaterials like fullerene and carbon nanotubes. All these features make ND as an attractive tool for both in vivo and in vitro applications [38, 50, 78, 79]. Conversely, there are many concerns and challenges related to their structure and nature toward their interactions with the living cells. Therefore, this situation requires many in- depth studies about the interaction nature between ND particles and the living cells. For example, some types of ND have a strong tendency to aggregate in specific medium which hindering and limiting their applications [48]. In fact, this aggregation tendency is related upon some synthesis techniques leads to produce ND particles with high dangling bonds on their surface such as detonation technique [48]. Hence the free electrons of the surfaces tend to form many functional groups and then these functional groups tend to form covalent bonds with the primary particle forming core- aggregates [48]. On the other side, the existing of sp2C around ND leads to bond these particles together into core aggregate [48].

Several methodologies have been carried for disintegration process, the most attractive one is the beads assisted sonication method which involving the double actions of shear force (induced by using zirconia beads) and cavitation effect (induced by ultrasound waves). The results shows colloid stability for one year after sonication for one hour [80]. While surface functionalization process has been recommended as an active way to reduce the aggregation size of ND. In fact there are several surface functionalization techniques, one of them involving generating specific surface radicals which then will act as a substrate used for synthesis of ND with carboxylic acid and dicarboxylic acid functionalization [48].

From a biocompatibility point of view, several studies have been demonstrated that ND toxicity can be varied and it is affected by surface chemistry of ND, cell- line type and the composition of treatment medium. In this field, the mitochondrial activity and the inflammatory activity of the cell have been used as toxicity indicator. So depending on these keys and the results of many clinical experiments, it was found that there is no manifestation of toxicity with ND dose (100 μg/ml concentration) after 24 hour of incubation period [81]. In addition, ND with high loading capacity, payload with high concentrations is allowed via using less delivery agents in association with the ability of releasing the cargo from the carrier in controlled manner, these two important features of ND developing the bio- applications of them either for small molecules delivery or for bio- technology product delivery [48].

In 2007, the suitability of ND particles as a delivery agent of doxorubicin hydrochloride (DOX) was studied by H. Huang et al. [81]. The study was based on the rationale that the surface carboxylic and hydroxylic groups of ND can interacts with the amine groups of DOX via ionic- forces when dispersing them in aqueous medium. The surface loading of DOX on ND particles was increased from 0.5 to 10 wt% via addition of 1% solution of sodium chloride to their aqueous dispersion, and the removal of salt favored the release of DOX. ND particles loaded with DOX were recommended to assemble in the form of loose- clusters, such that a certain amount of DOX adsorbed on the ND particle's surface resides within the cavity of the cluster [81].

This methodology of drug- entrapment in loose aggregates of ND particles could provide a feature by minimizing the systemic adverse effects of the naked- DOX. Thus, ND-based delivery systems could overcome the problem to the use of high concentrations of chemotherapeutic drugs in cancer treatments. In addition, the lower levels of cytotoxicity of the ND-DOX composites in mouse macrophages and human colorectal cancer cells compared with bare DOX in a 48-hour period could be beneficial in sustained drug release [81]. The potential of using ND particles as a targeted protein-delivery- vehicle was investigated in a pH-dependent system. By means of the physical- adsorption, ND particles achieved a considerable high surface loadings of bovine insulin about80% in pH-neutral water with a weight ratio of 1:4 of insulin:ND. Also, the aggregation properties of the insulin improved after interacting with ND particles. This propose that ND particles have the ability to facilitate the formation of a uniform-sized complex. Further, the release of insulin from the ND particle's surface was about 20- times higher when at a pH of about 10.5 than when in a neutral pH medium. Another advantage of using ND particles is the viability of cells, this effect had been observed with sodium hydroxide treated ND-insulin higher than what observed in neutral pH treated ND-insulin. Hence, the inherent and enhanced characteristics of ND particles make them as an active tool in drug delivery systems [48].

10. Conclusion

In summary, fullerene and nanodiamonds have been studied for drug delivery applications. Fullerenes and nanodiamonds are attractive allotropes in the carbon nanomaterials family. They can be synthesized with attractive properties in higher purity, higher surface homogeneity, and different surface functionalization and in controlled sizes that make them essential in nanomedicine fields. Utilization of fullerenes and nanodiamonds in drug delivery systems show higher advantages with enhanced targeted delivery and controlled drug release ability than other traditional strategies. But on the other side, further in-depth research about toxicity concerns are necessary in order to achieve the full advantages of utilization these nanomaterials in human body.

Author details

Basma H. Al-Tamimi* and Saad B.H. Farid
Department of Materials Engineering, University of Technology, Baghdad, Iraq

*Address all correspondence to: 130074@uotechnology.edu.iq

IntechOpen

© 2021 The Author(s). Licensee IntechOpen. This chapter is distributed under the terms of the Creative Commons Attribution License (http://creativecommons.org/licenses/by/3.0), which permits unrestricted use, distribution, and reproduction in any medium, provided the original work is properly cited.

References

[1] B S Murty, P Shankar, Baldev R, B B Rath, James M. Textbook of Nanoscience and Nanotechnology: Springer; 2013. 231 p. DOI: 10.1007/978-3-642-28030-6

[2] Luisa F, Duncan S. NANO TECHNOLOGIES: Principles, Applications, Implications and Hands-on Activities: Eurpean Commission Directorate- General for Research and Innovation Industrial technologies (NMP) program; 2013. 391 p. DOI: 10.2777/76945

[3] Qiong W, Wei-shou M, Yi-du Z, Han-jun G, David H: Mechanical properties of nanomaterials: A review. De Gruyter; Nanotechnol Rev. 2020; 9;259-273. DOI: 10.1515/ntrev-2020-0021

[4] Olga Z, Günter N: Carbon nanomaterials: production, impact on plant development, agricultural and environmental applications: Chem. Biol. Technol. Agric. Springer. 2016; 3: 17. DOI: DOI 10.1186/s40538-016-0070-8

[5] Ibrahim K, Khalid S, Idrees K: Nanoparticles: Properties, applications and toxicities. Arabian Journal of Chemistry. 2017; 12; 908-931. DOI: 10.1016/j.arabic.2017.05.011

[6] Seema T, R. M. Mehta: Fullerenes: An Introduction and Overview of Their ullerenes: An Introduction and Overview of Their Biological Properties. Indian Journal of Pharmaceutical Sciences. 2006; 13-19.

[7] Megha C, Anuptha P, Vianessa N, Daniel C, Devika C, Ronald H, Motahareh S, Sung K, Ashley K, Vesselin S, Mark S: Carbon Nanotube Sheet-Synthesis and Applications. Nanomaterials. 2020; 10; 2023. DOI: 10.3390/nano10102023

[8] Ruben M, Cristina G, Julio G, Felix Z: 2D materials: to graphene and beyond. Nanoscale. 2011; 3; 20. DOI: 10.1039/c0nr00323a

[9] Khalid M: Nanodiamond as a drug delivery system: Applications and prospective. Journal of Applied Pharmaceutical Science. 2011; 01(16); 29-39.

[10] Jing W, Zhongbo H, Jianxun X, Yuliang Z: Therapeutic applications of low-toxicity spherical nanocarbon materials. NPG Asia Materials; 2014; 6; 1-12. DOI:10.1038/am.2013.79

[11] P. S. Karthik, A. L. Himaja and Surya Prakash Singh: Carbon-allotropes: synthesis methods, applications and future perspectives: Review Articles. Carbon Letters. 2014; 15: 219-237. DOI: I:10.5714/CL.2014.15.4.219

[12] Thomas W, Clayton R, Jens O. Carbon Isotopes: Springer; 2018. p. 1-11. DOI: 10.1007/978-94-007-6238-1_44

[13] Noah P, Camille P, Andrey B. Carbon Isotopes as a Geochemical Tracer. In: Encyclopedia of Astrobiology: Springer; 2014. p. 1-6. DOI: 10.1007/978-3-642-27833-4_228-2

[14] Wojciech K, Sławomir D. Transition metal impurities in carbon-based materials: Pitfalls, artifacts and deleterious effects: Review article. Carbon. 2020. 168: 748-845

[15] TRENTON D H: DETERMINING THE sp2/sp3 BONDING CONCENTRATIONS OF CARBON FILMS [Thesis]. Saskatoon: University of Saskatchewan; 2005.

[16] Manijeh R. The carbon atom. In: The Mystery of Carbon. An introduction to carbon materials: IOP SCIENCE; 2020. p. 1-12. DOI: 10. 1088/2053-2563/ab35d1ch1

[17] E. Belenkov. Formation of Graphite Structure in Carbon Crystallites.

Inorganic Material. 2001; 37: 1094-1101. DOI: 10.1023/A: 1011601915600

[18] G. Smith. The allure, magic and mystery—A brief history of diamonds. The Journal of the South African Institute of Mining and Metallurgy. 2003

[19] John R. Amorphous carbon. In: Amorphous materials. 1996; 1:557-561

[20] LATHA K. S V SUBRAMANYAM. Structural and electrical properties of amorphous carbon–sulfur composite films. Bull. Mater. Sci. 2004; 27: 289-294

[21] Vasilios G, Jason A. P, Jiri T, Radek Z: Broad Family of Carbon Nanoallotropes: Classification, Chemistry, and Applications of Fullerenes, Carbon Dots, Nanotubes, Graphene, Nanodiamonds, and Combined Superstructures. Chemical Reviews. 2015; 115; 4744–4822. DOI: 10.1021/cr500304f

[22] Marta S, Ilona D, Anna J, Dariusz S. Graphene: One Material, Many Possibilities—Application Difficulties in Biological Systems. Journal of Nanomaterials. 2014; 2014: 11. DOI: 10.1155/2014/890246

[23] Ali E, Hadis D, Hamzeh K, Mohammad K, Nosratollah Z, Abolfazl A, Mozhgan A, Younes H, Sang W. Carbon nanotubes: properties, synthesis, purification, and medical applications. Nanoscale Research Letters. 2014; 9: 393-406. DOI: 10.1186/1556-276X-9-393

[24] Jeremy J. What is nanotechnology?. In: Nanotechnology; 2016. p. 1-18. DOI: 10.1016/B978-0-323-39311-9.00007-8

[25] Salisu N, Mohd Z, Zulkarnain Z, Nor A. Carbon-Based Nanomaterials/ Allotropes: A Glimpse of Their Synthesis, Properties and Some Applications. Materials. 2018; 11: 295-319

[26] Katharina K, Lorel M, Fabian S, Przemyslaw G, Leo G, Harry L. An sp-hybridized molecular carbon allotrope, cyclo[18]carbon. Science. 2019; 365: 1299-1301. DOI: 10.1126/ science.aay1914

[27] Igor L. Carbon (Graphene/ Graphite). In: General Classification of carbon allotropes: Springer; 2014. p. 7-32. DOI: 10.1007/978-94-007-7587-9_2

[28] Kunal B, Sourav P, Audrey G, Seth C, Silvia B, Stefano B, Massimo B, Alexander S, Bengt F. Biological interactions of carbon-based nanomaterials: From coronation to degradation. Nanomedicine: Nanotechnology, Biology, and Medicine. 2016; 12: 333-351. DOI: 10.1016/j. nano.2015.11.011

[29] Takuya Y, Kohei Y, Hiromi N, Tomoaki Y, Yasuo Y, Shin-ichi T, Yasuo T. Carbon Nanomaterials: Efficacy and Safety for Nanomedicine. Materials. 2012; 5: 350-363. DOI: 10.3390/ma5020350

[30] A Nimibofa, E Newton, A Cyprai and W Donbebe: Fullerenes: Synthesis and Applications. Journal of Materials Science Research. 2018; 7; 22-36. DOI:10.5539/jmsr.v7n3p22

[31] Tandabany C. Dinadayalane and Jerzy Leszczynski: Fundamental Structural, Electronic, and Chemical Properties of Carbon Nanostructures: Graphene, Fullerenes, Carbon Nanotubes, and Their Derivatives

[32] G. N. Churilov: Synthesis of Fullerenes and Other Nanomaterials in Arc Discharge. Fullerenes, Nanotubes and Carbon Nanostructures. 2008; 16; 395-403. DOI: 10.1080/15363830 802281641

[33] Lee C, Hao W, Stephen K: Fullerene formation during production of chemical vapor deposited diamond.

Applied PhysicsLetters. 1995; 66(4); 430-432. DOI: 10.1063/1.114046

[34] Boris K: Formation of fullerene in Laser ablation of graphite in vacuum. 1997.

[35] S Samal, S K Sahoo: An overview of fullerene chemistry. Bull. Mater. Sci. 1997; 20(2); 141-230.

[36] Paul W, Nathan K, Christopher L, Hendrickson, John P, Christopher P, Yusuke N, Yuki S, Hisanori S, Alan G,Harold W. Closed network growth of fullerenes. Nature communications. 2012; 3: 855. DOI: 10.1038/ncomms1853

[37] Boris I, Oxana V, Leonardo C. Synthesis Techniques, Properties, and Applications of Nanodiamonds. Synthesis and Reactivity in Inorganic, Metal-Organic, and Nano-Metal Chemistry. 2010; 40: 84-101. DOI: 10.3109/10799890903555665

[38] Emanuela T, Silvia O, Giacomo R, Teresa L,Mariglen A, Marco R, Maria L. Nanodiamonds: the ways forward. In: Proceedings of AIP Conference; 2015; 1667: p. 020001-020008. DOI: 10.1063/1.4922557

[39] V. V. Danilenko. INTRODUCTION: On the History of the Discovery of Nanodiamond Synthesis. Physics of the Solid State. 2004; 46: 595-599

[40] Luca B, Massimo C, Michele O, Antonio M. Nanodiamonds: Synthesis and Application in Sensing, Catalysis, and the Possible Connection with Some Processes Occurring in Space. Appl. Sci. 2020; 10: 4094-4122. DOI: 10.3390/app10124094

[41] L. Vandenbulcke, T. Gries, J. Rouzaud. Nanodiamonds in dusty low-pressure plasmas. Appl. Phys. Lett. 2009; 94: 044106. DOI: 10.1063/1.3075604

[42] Lucie B, Antonin B, Jana L, Lubica S, Stepan P, Alexander K: The Application of Nanodiamond in Biotechnology and Tissue Engineering: Diamond and Carbon Composites and Nanocomposites. InTech. Open. 2016; 60-88. DOI: 10.5772/63549

[43] Plotnikov V, Makarov S, Bogdanov D, Bogdanov A. The Structure of Detonation Nanodiamond Particles. In: Proceedings of AIP Conference. Mechanics, Resource and Diagnostics of Materials and Structures (MRDMS-2016); 2016. p. 040045-1-040045-4

[44] Jae L H, Yong S Y, Dong H L: Thermal Oxidative Purification of Detonation Nanodiamond in a Gas-Solid Fluidized Bed Reactor. Korean Chem. Eng. Res. 2018; 56(5); 738-751. DOI: 10.9713/kcer.2018.56.5.738

[45] Vitaly I, Hiro-o H, Eiji O, Vladimir E, Igor K, Bastian J, Olga L, Boris Z, Chandra P, Huan-Cheng C. Carbon structure in nanodiamonds elucidated from Raman spectroscopy. Carbon. 2017; 121: 322-329. DOI: 10.1016/j.carbon.2017.06.012

[46] Aditi D, Mesut K, Yao Fu, Albert C. Surface structure and properties of functionalized nanodiamonds: a first-principles study. In: Proceedings of IOP on Nanotechnology; 2011. p. 065706

[47] J.C. Arnaul, H.A. Girard. Hydrogenated nanodiamonds: Synthesis and surface properties. Current Opinion in Solid State and Materials Science. 2016; 21: 10-16. DOI: 10.1016/j.cossms.2016.06.007

[48] Randeep K, Ildiko B. Nanodiamonds as novel nanomaterials for biomedical applications: drug delivery and imaging systems. International Journal of Nanomedicine. 2013; 8: 203-220. DOI: 10.2147/UN.S37348

[49] Ajay K, Pin A, Albert X, Boyi H, Yoke K, R. Mohan. Formation of nanodiamonds at near-ambientconditions via microplasma dissociation of ethanol vapour. NATURE COMMUNICATIONS. 2013; 2: 2618. DOI: 10.1038/ncomms3618

[50] Vadym N, Olga S, Dean H, Yury G. The properties and applications of nanodiamonds. NATURE NANOTECHNOLOGY. 2011. DOI: 10.1038/NNANO.2011.209

[51] Awadesh K M, Joana C M, Shlomo Z R, Sandip B: Detonation Nanodiamond Seeding Technique for Nucleation Enhancement of CVD Diamond – Some Experimental Insights. Advances in Ceramic Science and Engineering. 2014; 3; 36-45. DOI: 10.14355/acse.2014.03.005

[52] Awadesh K M, S R Binu, L N Satapathy, Chandrabhas N, M D Motin S, S A Shivashankar, S K Biswas: Effect of substrate roughness on growth of diamond by hot filament CVD. Bull. Mater. Sci. 2010; 33 (3); 251-255.

[53] Awadesh K M, Sandip B, Kalyan S P, Nandadulal D: Synthesis and charaterization of freestanding diamond coating. Indian Journal of Engineering and Materials Science. 3013; 20; 522-532.

[54] Awadesh K M, Nandadulal D, Shirshendu C, Ashok K M, Jiten G, Manju U, Sandip B, Vamsi K B: Characterizations of microwave plasma CVD grown polycrystalline diamond coatings for advanced technological applications. Processing and Application of Ceramics. 2014; 8 (2); 69-80. DOI: 10.2298/PAC1402069M

[55] James E, Anirudha V. The CVD of Nanodiamond Materials. Chem. Vap. Deposition. 2008; 14:145-160. DOI: 10.1002/cvde.200700037

[56] Maria L, Silvia O, Marco R, Emanuela T. Nanodiamonds for field emission: state of the art. Nanoscale. 2015; 7: 5094-5114. DOI: 10.1039/c4nr07171a

[57] A. Miotello, R. Kelly. Laser-induced phase explosion: new physical problems when a condensed phase approaches the thermodynamic critical temperature. Appl. Phys. A. 1999; 69: s67-s73. DOI: 10.1007/s003399900296

[58] A. Panich, A.I. Shames, B. Zousman, O. Levinson. Magnetic resonance study of nanodiamonds prepared by laser-assisted technique. Diamond & Related Materials. 2013; 23:150-153. DOI: 10.1016/j.diamond.2011.12.047

[59] Snigdha R, Mitun D, Awadesh K M, Vamsi K B: Laser melting of titanium-diamond composites: Microstructure and mechanical behavior study. Materials Letters. 2016; 178; 284-287. DOI: 10.1016/j.matlet.2016.05.023

[60] Reza M, S Siamak A T: The effect of ultrasonic irradiation on the structure, morphology and photocatalytic performance of ZnO nanoparticles by sol-gel method. Ultrasonics – Sonochemistry. 2017; 39; 504-510. DOI: 10.1016/j.ultsonch.2017.05.012

[61] Jin Ho B, Kenneth S: Applications of Ultrasound to the Synthesis of Nanostructured Materials. A dv. Mater. 2010; 22: 1039-1059. DOI: 10.1002/adma.200904093

[62] Basma H, Iman I, Haitham M. Synthesis and characterization of nanocrystalline diamond from graphite flakes via a cavitation-promoted process. Heliyon. 2019; 5: 1-5. DOI: 10.1016/j.heliyon.2019.e01682

[63] A.Kh. Khachatryan, S.G. Aloyan, P.W. May, R. Sargsyan, V.A. Khachatryan, V.S. Baghdasaryan:

Graphite-to-diamond transformation induced by ultrasound cavitation Diamond and Related Materials. 2008; 17: 931-936. DOI: 10.1016/j.diamond.2008.01.112

[64] Ali R, Derek J. Preparation of nanodiamonds from carbon nanoparticles at atmospheric pressure. Chem. Commun. 2015; 51: 5594—5597. DOI: 10.1039/c5cc00233h

[65] E. G. Gamaly, A. V. Rode, B. Luther-Davies. Ablation of solids by femtosecond lasers: Ablation mechanism and ablation thresholds for metals and dielectrics. Phys. Plasmas. 2002; 9: 949-957. DOI: 10.1063/1. 1447555

[66] Francisco C, Ricardo E, Jefferson B, Raul O, Nilson D, Narcizo M. Synthesis of diamond-like phase from graphite by ultrafast laser driven dynamical compression. Scientific Reports. 2015; 5: 1-6. DOI: DOI: 10.1038/srep11812

[67] Surendiran A, Selvarajan S, Suresh C P, C. Adithan. Novel applications of nanotechnology in medicine. INDIAN J MED RES. 2009; 13: 689-701

[68] Abu Mohammad A, Mohammad K, Md. Abdul Khaleque. The Application of Nanotechnology in Medical Sciences: New Horizon of Treatment. Am. J. Biomed. Sci. 2017; 9: 1-14. DOI: 10.5099/aj170100001

[69] Frank A, Eric M, Robert L, Omid C. Handbook of Experimental Pharmacology 197: Nanoparticle Technologies for Cancer Therapy: Springer; 2010. p. 14-21. DOI: DOI 10.1007/978-3-642-00477-3_2

[70] Biswajit M, Niladri S, Ruma M, Priyanka B, Pranab J, Paramita P. In: Current Status and Future Scope for Nanomaterials in Drug Delivery. Handbook of Application of Nanotechnology in Drug Delivery: IntechOpen; 2014. p. 525-544. DOI: 10.5772/58450

[71] Wim H, Paul J. Drug delivery and nanoparticles: Applications and hazard. International Journal of Nanomedicine. 2008; 3: 133-149

[72] Ramesh R, R N Okigbo, Madhusoodhan S, SangeetaC. Nanotechnology importance in the pharmaceutical industry. African Journal of Pure and Applied Chemistry. 2008; 2: 027-031

[73] Ajay K, FULLERENES FOR BIOMEDICAL APPLICATIONS. Journal of Environmental and Applied Bioresearch. 2015; 3: 175-191

[74] Ülkü A, Suna T, Zekerya D. Recent pros and cons of nanomaterials in drug delivery systems. INTERNATIONAL JOURNAL OF POLYMERIC MATERIALS AND POLYMERIC BIOMATERIALS. 2019; DOI.org/10.1080/00914037.2019.1655753

[75] Robert D, Fullerenes for Drug Delivery. In: Nanomaterials for Electrical Energy Storage Devices: Editors: Bharat Bhushan, Encyclopedia of Nanotechnology. 12 edition Springer. 10.1007/978-90-481-9751-4_58

[76] Tatiana Y, Alexander S, Balaji S, Brian E, Vernon K. Lon J. A Fullerene-Paclitaxel Chemotherapeutic: Synthesis, Characterization, and Study of Biological Activity in Tissue Culture. J. AM. CHEM. SOC. 2005; 127: 12508-12509

[77] Rania B, Rainer M, Muhammad N, Matthias R, Zoltan S, Christian W, Günther K. Medicinal applications of fullerenes. International Journal of Nanomedicine. 2007; 2: 639-649

[78] Anuradha Jana, Nandadulal Dandapat, Somoshree Sengupta, Vamsi Krishna Balla, Rajnarayan Saha, Awadesh Kumar Mallik: Human

osteoblast like MG63 cell and mouse fibroblast NIH3T3 cell viability study on the nucleation side of CVD grown polycrystalline diamond coatings. Trends in Biomaterials and Artificial Organs. 2015; 3; 211-216.

[79] Anke K: The structure and reactivity of nanoscale diamond. J. Mater. Chem. 2008; 18: 1485-1492. DOI: DOI: 10.1039/b716673g

[80] Masaki O, Masayasu I, Makoto T, Fumiaki K, Anke K, Eiji O. Preparation and Behavior of Brownish, Clear Nanodiamond Colloids. Adv. Mater. 2007; 19: 1201-1206. DOI: DOI: 10.1002/adma.200601452

[81] Houjin H, Erik P, Eiji O, Dean H. Active Nanodiamond Hydrogels for Chemotherapeutic Delivery. Nano Lett. 2007; 7: 3305-3314. DOI: 10.1021/nl071521o

Chapter 6

Nano-Sized Minerals from Lower Cretaceous Sandstones in Israel Observed by Transmission Electron Microscopy (TEM)

Nurit Taitel-Goldman and Vladimir Ezersky

Abstract

Fine fraction in quartz arenite sandstones from Lower Cretaceous Hatira formation in Israel was observed by Transmission electron microscope (TEM). Samples were collected from Hatira and Ramon craters located in southern part of Israel and from Manara cliff from the northern part of Israel. The additional phases cause yellow, red, dark red and dark violet colors of the layered sandstones. The motivation was to identify the minerals of the fine factions that cause the variations in the colors. The minerals observed were clay minerals, mainly kaolinite ($Al_4Si_4O_{20}(OH)_8$), some illite ($K_{0.65}Al_{2.0}[Al_{0.65}Si_{3.35}O_{10}](OH)_2$) and smectite. Iron oxides were goethite (FeOOH) and hematite (Fe_2O_3), Titanium-iron oxides observed was ilmenite ($FeTiO_3$), and Titanium-oxides were rutile (TiO_2), and anatase (TiO_2). Sulphates observed were jarosite ($KFe_3(SO_4)_2(OH)_6$) and alunite ($KAl_3(SO_4)_2(OH)_6$). Some of the hematite was formed by recrystallization of goethite. Ilmenite disintegrated into small iron oxides mainly hematite. Euhedral to sub-hedral rutile (TiO_2) and anatase (TiO_2) were preserved in clay-minerals. Crystals of alunite and jarosite were observed in sandstones in both craters. They probably crystallized due to some transgression of the Thetis Sea.

Keywords: TEM, sandstones, clay minerals, Fe-oxides, Ti-oxides, sulphates

1. Introduction

During the Lower Cretaceous, siliciclastic sediments were deposited on terrestial terrestrial partly lacustrine environments [1–3]. The sandstones are quartz (SiO_2) arenite with rounded quartz grains, clay minerals and siltstones. Lower Chemical weathering of the Pan-African continental basement favored silicate weathering, particularly a warm and humid climate, low relief and low sedimentation rates which prevailed over large tracts of Gondwana in the aftermath of the Pan-African orogeny [4]. Cretaceous sandstone originates from Paleozoic sandstones, the first-cycle quartz-rich sandstones [5]. The stratigraphic cycles showed progradational – retrogradational trends due to small global sea-level rises which inundated the continent for short periods and fossils were found [6]. In Lebanon similar sandstones were deposited during the Lower Cretaceous [7].

The composition of the sandstones was 85–95% quartz indicating well-sorted sandstone. Sedimentological observations suggest that the Chouf Formation deposited in fluvial, coastal plain and deltaic environments. The origin of the sandstone is from recycling of Paleozoic sandstones. Sandstones of the Lower Cretaceous were deposited and covered by sediments that were deposited after the Thetis Ocean covered the area during the Upper Cretaceous, hence, hard carbonate stones, limestones, dolomites and marls precipitated, overlying the sandstones. At the end of the Mesozoic era, closure of the Thetis Sea yielded formation of the Syrian arc with monoclines of the Hatira and Ramon areas. After regression of the Thetis Sea during the Oligocene, an erosion surface truncated the hard carbonate rocks of the fold's crest and exposed the underlying friable Lower Cretaceous sandstone. For crater (Makhtesh) Hatira, such an erosion base level formed locally and for a limited period in the Early Miocene. The opening of the Ramon crater started in the Early Pliocene, when the Syrian Arc Fold Belt was uplifted and arched and the Dead Sea Rift was established as a deep intercontinental erosion base in the east [8]. In the northern part of Israel, the Lower Cretaceous sandstones were exposed close to the Dead Sea transform fault and uplift of the Manara cliff and formation of Hula valley.

Composition and morphology of the nano-crystals that coat quartz grain reflect the environmental conditions in which they were formed as primary or secondary forms [9].

Figure 1.
a. Geomorphological map of Israel. In the northern part is Manara cliff and in the southern part Hatira and Ramon craters. b. Sandstones exposed in the Hatira crater and a sample of red and yellow layers. c. Sandstones exposed in the Ramon crater and a sample with dark violet, red and yellow-red thin layer. d. Sandstones in the Manara cliff.

In this paper, we present Transmission Electron Microscopic (TEM) observation of the nano-sized minerals from Lower Cretaceous sandstones exposed in the Hatira and Ramon craters in southern Israel, and in the Manara cliff in the northern part of Israel (**Figure 1**).

The main motivation of this research is to identify nano-crystals in sandstone that cause the variability in their colors. By using TEM detrital or authigenic phases can be identified.

2. Sampling and methods

The samples that were collected were separated according to their colors: dark violet, violet, dark red, red and yellow (**Table 1**). Sand stone were put for few minutes into an ultrasonic bath, then the fine fraction was separated from the quartz grains and freeze-dried. Each sample was suspended in distilled water and placed on Cu supported holey carbon film and dried. Nano-sized (5–200 nm) particles were checked with Transmission Electron Microscopy (TEM) using a JEOL JEM-2100F analytical TEM operated at 200 kV, equipped with a JED-2300 T Energy Dispersive Spectrometer (EDS) for microprobe elemental analyses. All chemical analyses were obtained by point analyses with a beam width of 1 nm and are presented as atomic ratios. JEOL Analytical Station software, based on the Cliff-Lorimer ratio technique, with an accuracy of ~5%, was used for the calculations. The CuKa line was used to calibrate the spectrometer. Energy-filtered TEM (EFTEM) experiments were performed using a Gatan image filter. The titanium L-edge (456 eV), silicon L-edge (99 eV) and iron L-edge (708 eV) were used for elemental mapping using the three-window method.

No.	Sample	Color	Location	Minerals identified with HRTEM
1	GNH4 YR	Yellow-red	Hatira crater, east	Kaolinite, illite, hematite, goethite
2	GNH4 DR	Dark red	Hatira crater, east	Quartz, clays, jarosite, anatase, goethite, calcite, ilmenite
3	GNH6 R	Red	Hatira crater, east	Kaolinite, ilmenite, hematite, quartz
4	GNR1 V	Violet	Ramon crater northern cliff highway 40	Anatase, kaolinite illite, smectite, goethite, magnetite, maghemite
5	GNR1 Y	Yellow	Ramon crater northern cliff highway 40	Hematite, illite, rutile, quartz
6	GNR2 DV	Dark violet	Ramon crater northern cliff highway 40	Hematite, goethite, kaolinite alunite, apatite
7	GNR2 R	Red	Ramon crater northern cliff highway 40	Goethite, hematite, anatase, illite, kaolinite
8	GNR2 VR	Violet, red	Ramon crater northern cliff highway 40	Hematite, goethite, kaolinite anatase
9	GNR3 R	Red	Ramon crater northern cliff highway 40	Goethite, hematite, kaolinite, illite
10	GNR5 R	Red	Ramon crater northern cliff highway 40	Goethite hematite, anatase, kaolinite, illite, smectite
11	GNKS1	Red	Manara cliff near Kiriat Shmona	Kaolinite, ilmenite, apatite, goethite, quartz

Table 1.
Location, colors and mineral composition in the fine fractions in sandstones from Hatira crater, Ramon crater, and Manara cliff in Israel

Crystalline phases were identified, using Selected Area Electron Diffraction (SAED) in the TEM. With this method, a very high-energy electron beam (200 kV) transmits through the sample and the d-values obtained enabled their identification. The accuracy of the d-values determination was better than 0.005 nm. The smallest area of SAED was 100 nm. If a good lattice image was obtained in very small particles by High Resolution Transmission Electron microscopy, the use of Fast Fourier Transformation (FFT) enabled the identification of the minerals, using the program Digital Micrograph (Gatan).

3. Results

3.1 Hatira crater sandstones

Samples from the Hatira crater were collected at the eastern side of the crater. The colors of the sandstones vary, with thin layers of red sandstone or dark red close to clay layers. Usually the yellow layers are thicker.

In the sample: yellow-red (GNH4 YR), from the Hatira crater (**Figure 2**), a cluster of goethite (FeOOH) with clay minerals, mainly kaolinite ($Al_4Si_4O_{20}(OH)_8$), was observed, and a cluster of hematite (Fe_2O_3) shows the previous morphology of goethite, indicating that hematite was formed by recrystallization of goethite preserving the acicular morphology of goethite. Crystal size of hematite is around 15 nm. Small crystals (25-50 nm) with Ti impurity might reflect ilmenite ($FeTiO_3$) disintegration. The dominance of goethite crystals contributes to the yellow color and hematite contributes to the red color. The recrystallization of goethite into hematite might have happened due to thermal transformation, as the sandstones were covered by younger layers.

Sample GNH4 DR (**Figure 3**) had a dark red color and the minerals identified were clay minerals, small crystals of calcite ($CaCO_3$), jarosite ($KFe_3(SO_4)_2(OH)_6$), goethite, ilmenite, and anatase (TiO_2). The image obtained by TEM shows clay

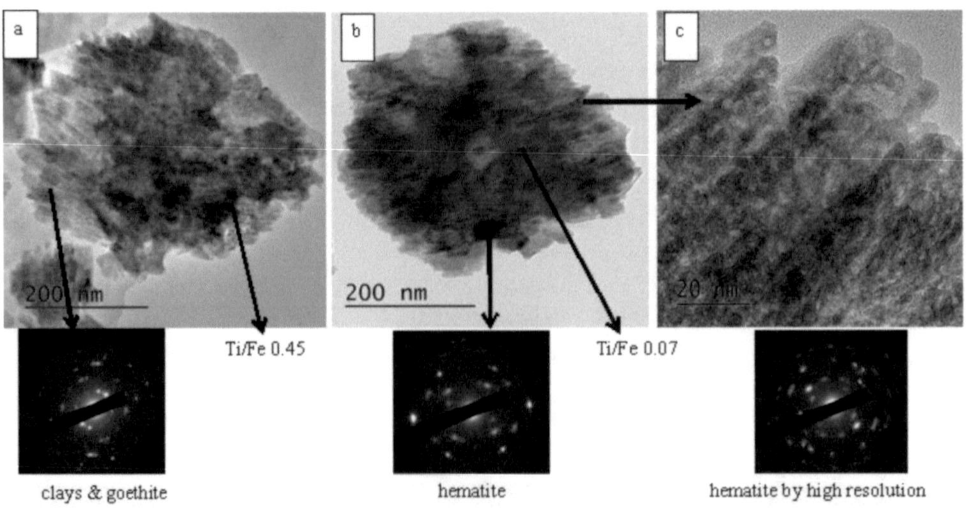

Figure 2.
GNH4 YR images of yellow-red sandstone from the Hatira crater with electron diffractions. Minerals observed were: a. cluster of clay minerals (illite ($K_{0.65}Al_{2.0}[Al_{0.65}Si_{3.35}O_{10}](OH)_2$) and kaolinite, with goethite; electron diffractions of goethite 0.42–0.43 nm in the inner circle. b. Cluster of hematite; electron diffractions of hematite 0.36–0.37 0.268–0.269 nm. c. Tiny crystal of hematite high resolution of hematite formed by recrystallization of goethite preserving the initial phase of goethite. Small crystals had Ti/Fe ratios 0.07 and 0.45. Electron diffractions were 0.36 and 0.27 nm.

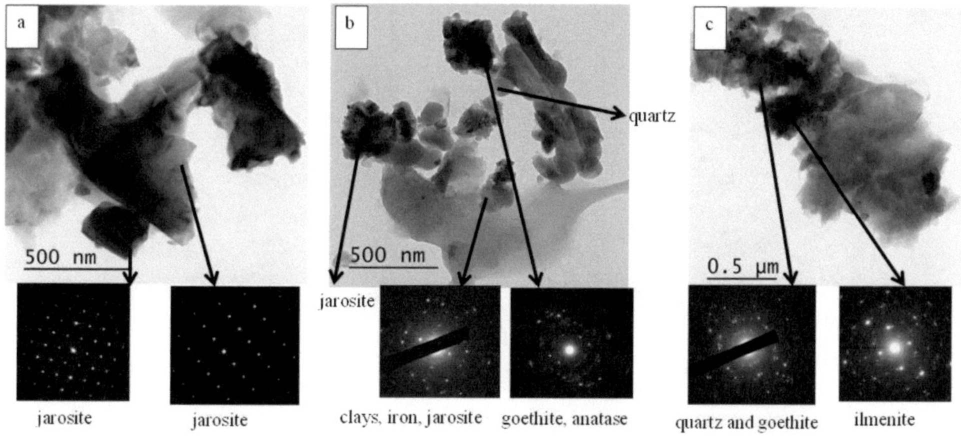

Figure 3.
GNH4 DR dark red sandstone from the Hatira crater. Minerals observed were clay minerals, jarosite goethite, anatase and ilmenite. a. Jarosite with point analyses (%atom) in the lower left side yielded Al-27.28, S-46.69 K-13.91 and Fe-12.12; in the upper part Al-24.38, S-52.13, K-9.32 and Fe-14.17. Electron diffractions at the left side were 0.59 nm, 0.57 nm and 0.28 nm at the central part the electron diffractions were 0.506 nm, 0.3108 nm and 0.3012 nm. b. Minerals observed were clay minerals, jarosite, goethite, anatase with atomic ratio Fe/Ti 0.85. Electron diffractions at the upper part of the images yielded for goethite 0.42 nm and for anatase 0.35. c. Minerals observed were quartz, and ilmenite with electron diffraction of 0.36 nm.

minerals with clusters of anatase and goethite. Jarosite appears with euhedral or sub-hedral morphology, indicating that it was formed in the sandstone. The crystal size of jarosite is 100-200 nm. Ilmenite crystal had Fe/Ti 0.85 and the crystal's size is $\approx$200 nm.

Sample GNH6 R (**Figure 4**) was collected from a thin red layer and vein within a clay layer in the sandstone. Ilmenite was found as a cluster of small particles surrounded by clay minerals. The ilmenite point analysis had Fe/Ti 0.38, indicating that a larger crystal disintegrated and was preserved within the clay layer. EFTEM RGB color shows disintegration of the ilmenite mineral. Usually ilmenite is resistant to weathering processes and it probably arrived at the area along with the quartz grains. At the lower part of the image hematite formed a layer, causing the red color of the sandstone. Recrystalliztion of ilmenite into hematite was found also in sands along the Mediterranean coast. The precursor and the recrystallized tiny hematite crystals remain close to each other since they are all kept within clay minerals [10].

3.2 Ramon crater sandstones

Sandstones are exposed at the northern cliff of the Ramon crater close to highway 40. Volcanic eruption occurred in the area during the Lower Cretaceous, forming basanite flows and paleosol between the volcanic flows [11]. The samples presented were collected above the basanitic flows. Lower cretaceous sandstones in the Ramon crater have various colors: dark violet, violet, dark red, red and yellow (**Figure 1c**).

The violet sample GNR1V (**Figure 5**) had kaolinite clusters, euhedral crystals of anatase (50-100 nm), magnetite (Fe_3O_4) and maghemite (Fe_3O_4) that was probably formed by magnetite oxidation. Kaolinite crystals had euhedral morphology, indicating that they crystallized in the area. Similar euhedral morphology of some of the anatase and rutile (TiO_2) crystals indicates that they might also have been formed in the sandstone by recrystallization of other phases. Another option is that the euhedral morphology of the Ti-oxides results from their resistance to weathering. In sample GNR1 V, anatase crystals reached a size of 90 nm and Fe/Ti atomic ratios were 0.04–0.26.

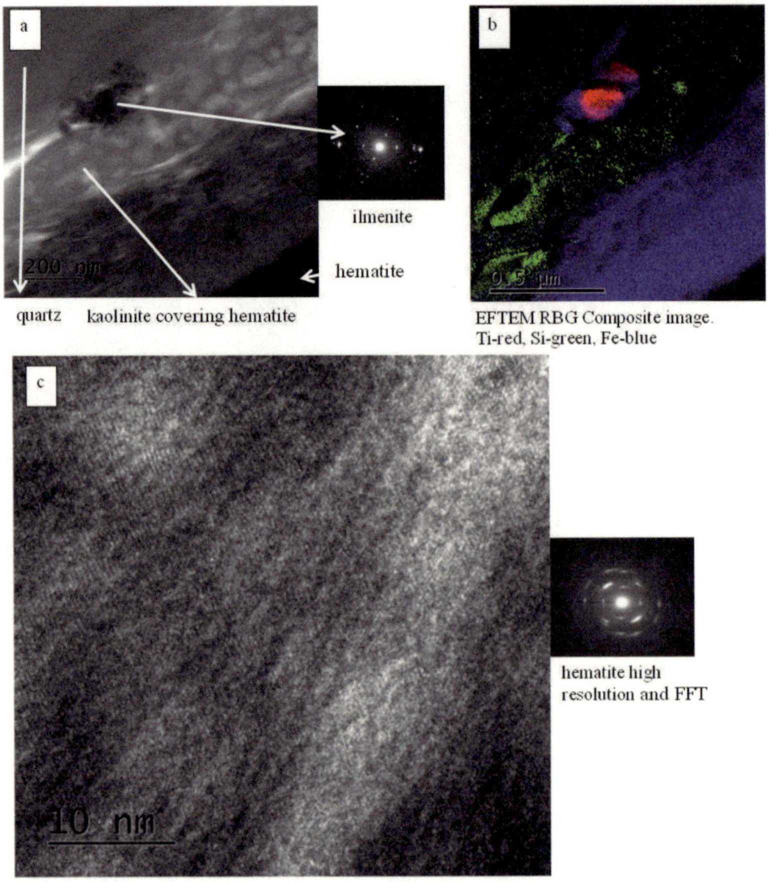

Figure 4.
GNH6 R red sandstone from the Hatira crater: a. cluster of ilmenite with Fe/Ti 0.38 in a kaolinite layer covering hematite layer. The upper gray layer is part of a quartz grain. b. Composition of phases observed on the left side. c. High resolution image of hematite with FFT.

Sample GNR1Y (**Figure 6**) presents additional phases of a yellow sandstone. Quartz grain was observed surrounded by a cluster of illite ($K_{0.65}Al_{2.0}[Al_{0.65}Si_{3.35}O_{10}](OH)_2$), hematite and a euhedral rutile crystal (250 nm) with Fe/Ti 0.03 atomic ratio. Hematite tiny crystals <100 nm were also observed close to the quartz grain and on the clays.

In sample GNR2 DV dark violet sandstone from the Ramon crater (**Figure 7**), the main clay mineral is kaolinite forming clusters with hematite and alunite ($KAl_3(SO_4)_2(OH)_6$). The largest crystal size of hematite is around 300 nm, contributing to the violet color of the sandstone. Point analysis in % atom of alunite yielded: Al 6.00–6.37, S 2.04–2.34, K 0.29–0.32 and Fe 0.01. Alunite was probably precipitated from acid rain or evaporation of sea water that might have entered the area during the Lower Cretaceous. Tiny crystals of goethite formed clusters with clays.

Sample GNR2 R from red sandstone in the Ramon crater (**Figure 8**) had clay minerals, illite and kaolinite with anatase, goethite and hematite. Anatase (~50 nm) had euhedral morphology, indicating that it might have crystallized in the sandstone. The goethite crystals had acicular morphology (~150 nm); some of the goethite partially recrystallized into hematite preserving the initial acicular goethite morphology. Tiny crystals of hematite were also observed on goethite crystals. A high resolution image was taken from a cluster of hematite preserving outer goethite morphology.

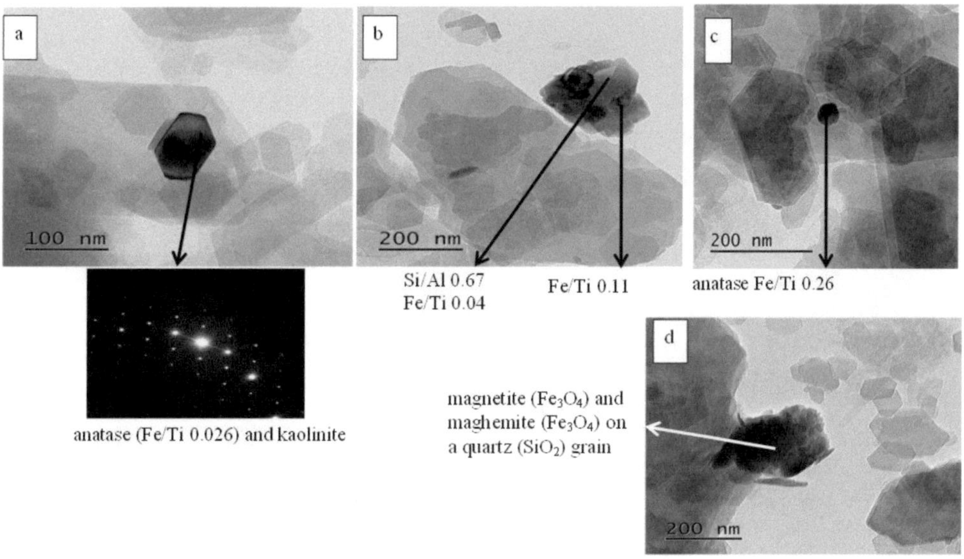

Figure 5.
GNR1 V violet sandstone from Ramon crater. Minerals observed: a, b, c: Anatase with electron diffraction at 0.36 nm and kaolinite. d: Magnetite (Fe_3O_4) and maghemite (Fe_3O_4) on a quartz (SiO_2) grain, kaolinite at the right side.

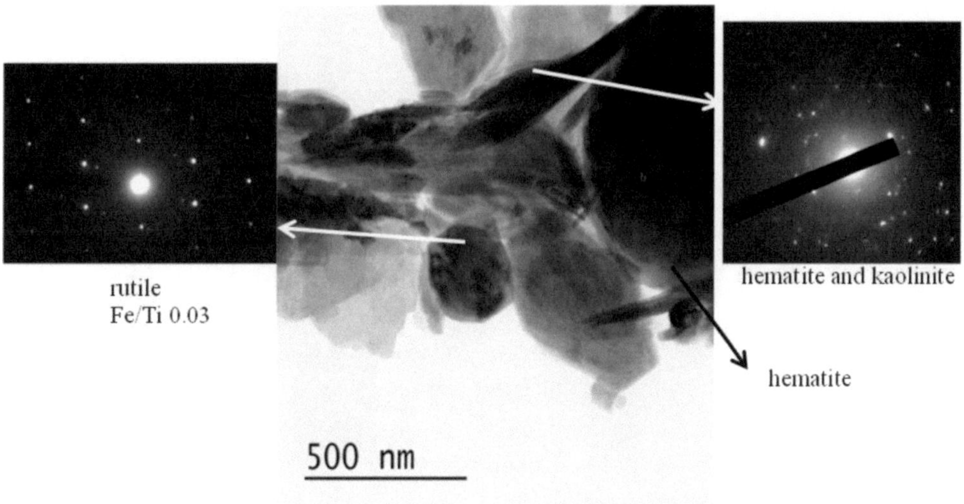

Figure 6.
GNR1 Y. image of yellow sandstone from Ramon crater with electron diffractions. The minerals observed were rutile, hematite illite and kaolinite. Quartz (SiO_2) grain is on the right side.

Sample GNR2 VR (**Figure 9**) is from violet-red sandstone from the Ramon crater. The minerals identified were kaolinite, goethite, hematite and anatase. High resolution of the sample yielded morphology of euhedral acicular crystals that were the initial stage of goethite, and the inner morphology was of small crystals of hematite. Goethite has been transformed into hematite preserving the acicular morphology. The cluster of hematite includes the clay mineral, kaolinite. Euhedral anatase crystal (~60 nm) was observed close to clay minerals with impurity of Fe/Ti 0.027 atomic ratio.

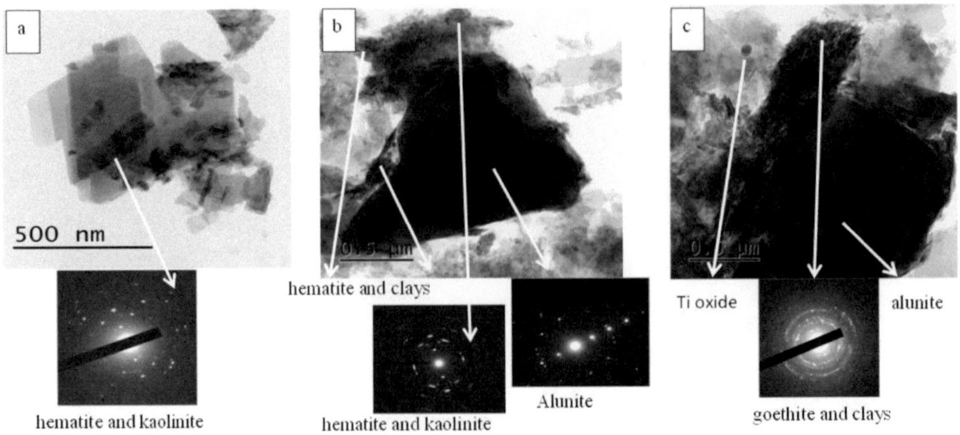

Figure 7.
GNR2 DV image of dark violet sandstone from Ramon crater. Minerals observed were: a. kaolinite and hematite; b. kaolinite, hematite with electron diffraction at 0.37 nm and at the lower part alunite with electron diffraction at 0.5 nm; c. small crystal of Ti-oxide, cluster of goethite with electron diffraction 0.42 nm and alunite crystal on the right side.

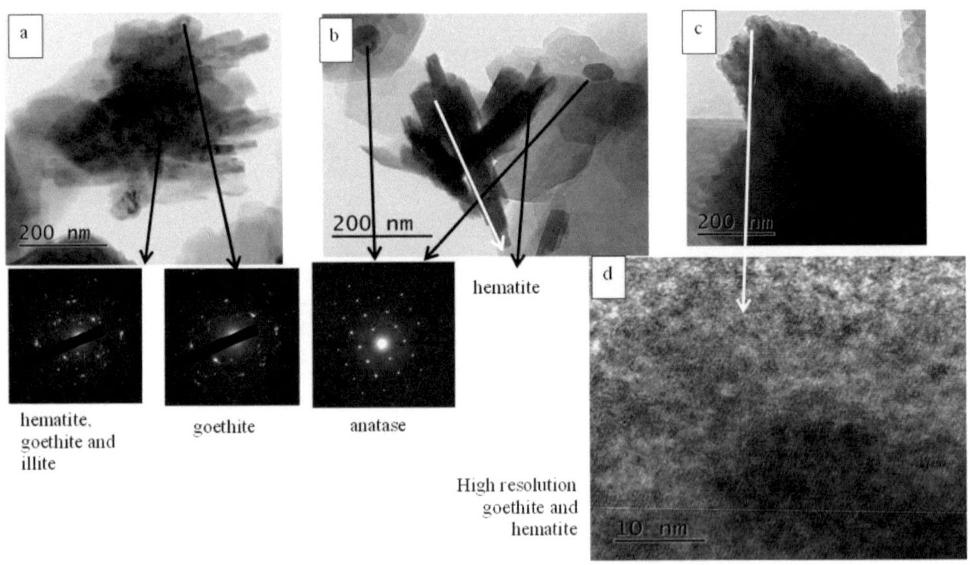

Figure 8.
GNR2 R red sandstone from Ramon crater. Minerals observed: a. goethite and hematite on illite and kaolinite; b. hematite and anatase; c. goethite and tiny crystal of hematite; d. high resolution of goethite and hematite.

Sample GNR3 R (**Figure 10**) red sandstone from the Ramon crater had clusters of goethite with kaolinite, illite and smectite. Small crystals (<50 nm) that were observed on the clays had impurity of Ti with the atomic ratio of Ti/Fe 0.027. These crystals might result from disintegration of ilmenite.

Sample GNR5 R (**Figure 11**) red sandstone from the Ramon crater had clay minerals, goethite crystals and anatase (TiO_2). The size of the goethite crystals reached 800 nm.The size of euhedral anatase crystal was 200 nm and impurity of iron was Fe/Ti 0.04 atomic ratio. From a dark field image obtained, tiny crystals of anatase are located in the upper part of the image.

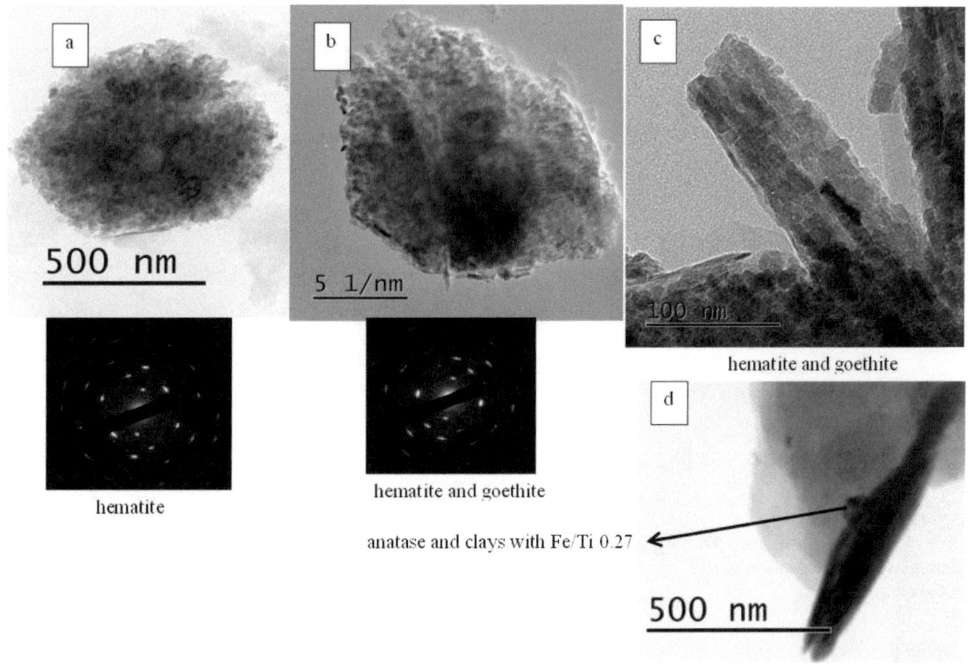

Figure 9.
GNR2 VR violet-red sandstone from Ramon crater. Minerals observed: a. a cluster of hematite with electron diffraction 0.367–0.374 nm; b. hematite electron diffraction 0.367–0.377 nm and goethite electron diffraction 0.42 nm; c. high resolution of hematite preserving initial crystals of goethite electron diffractions 0.365–0.375 nm; d. anatase (TiO_2) with iron impurity and clays.

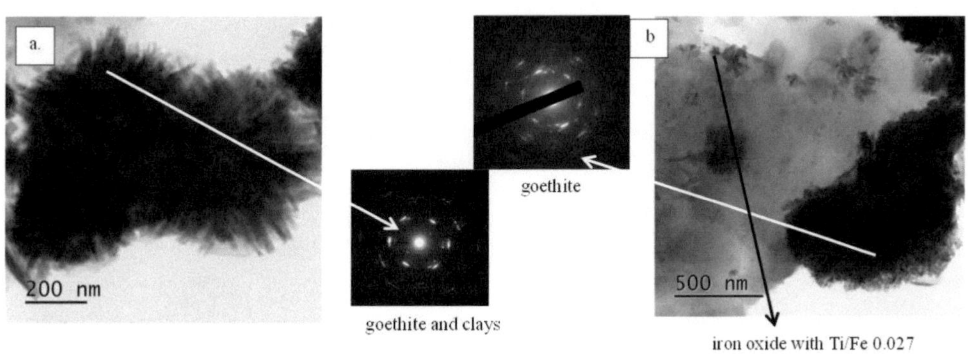

Figure 10.
GNR3 R red sandstone from Ramon crater: a. Kaolinite, smectite and illite with electron diffractions of 0.45nm and goethite. b. Lay minerals and goethite electron diffraction 0.41 nm.

3.3 Sandstone on Manara cliff

Sandstones of Lower Cretaceous were exposed in the northern part of Israel due to uplift along the Dead Sea transform fault. The main phases identified in the fine fraction were clay minerals, mainly kaolinite, ilmenite and hematite (**Figure 12**). The size of ilmenite crystals varies between 1000 nm and 50 nm. Most of the ilmenite crystals disintegrate into small crystals and Fe/Ti ratios decrease. The tiny ilmenite crystals are preserved in clays covering a quartz grain. Hematite tiny crystals that result from disintegration of ilmenite and recrystallize into hematite are preserved within clay minerals.

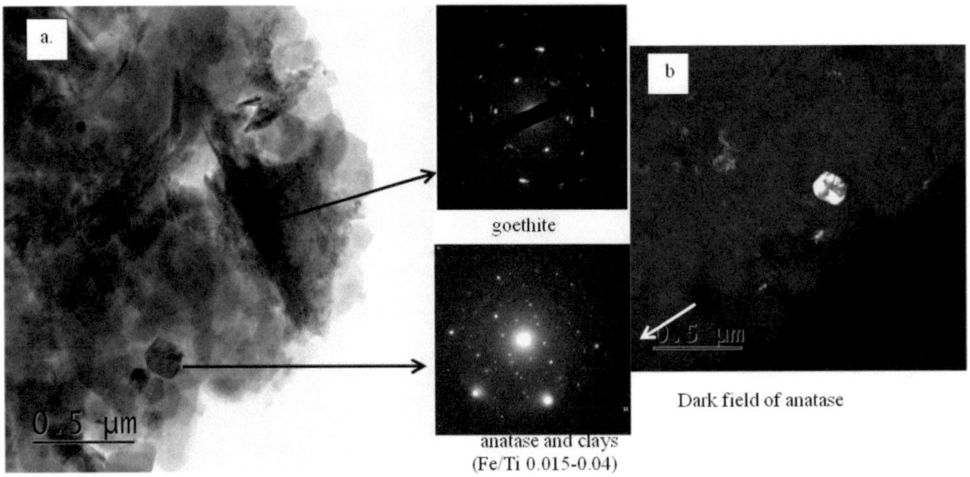

Figure 11.
GNR5 R red sandstone from Ramon crater. a. Minerals observed with electron diffraction were: anatase and goethite surrounded by clays. b. Dark field image showing larger anatase crystal and small crystals of anatase in the area.

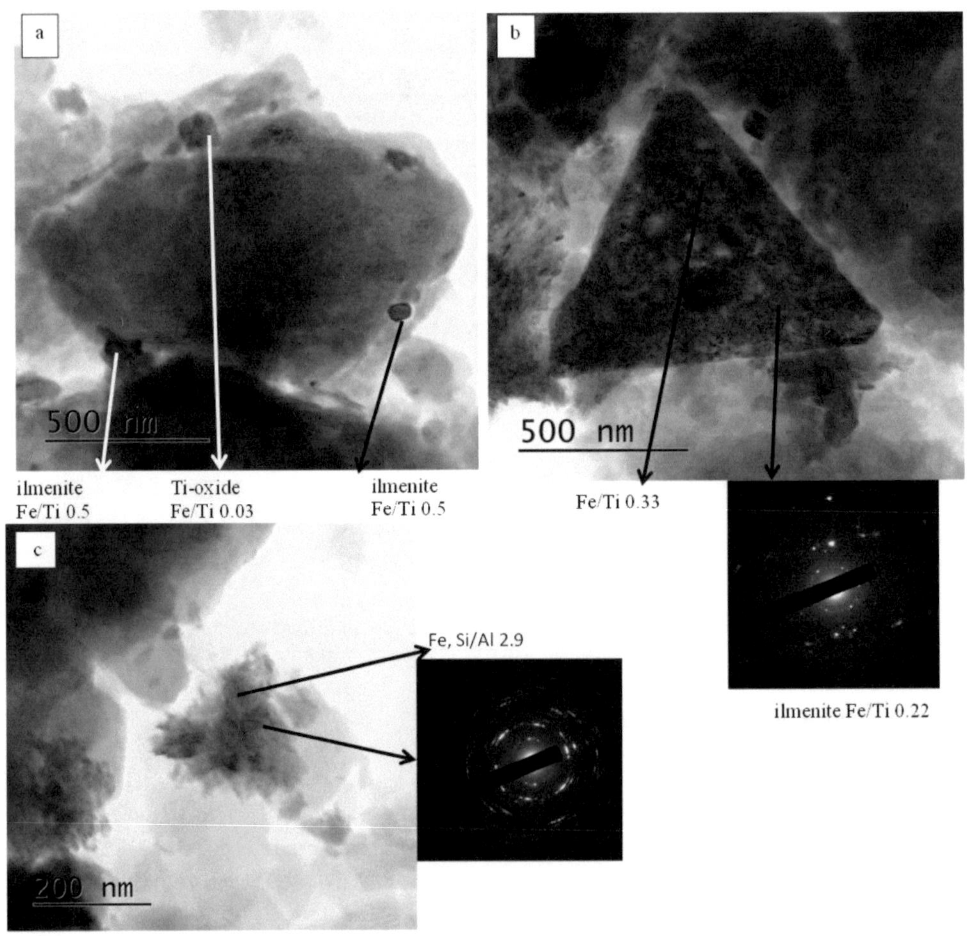

Figure 12.
GNKS1 sandstones exposed in Manara cliff. Minerals observed: a. clays and ilmenite; b. ilmenite electron diffraction 0.255 nm; c. cluster of hematite with electron diffraction of 0.37–0.38 nm and clays over a small quartz (SiO_2) grain on the upper left side quartz grain and clays.

4. Discussion

The Lower Cretaceous sandstones are mainly quartz arenite with rounded quartz grains that were separated from the fine fraction. These sandstones originated from Paleozoic sandstones that were the first-cycle quartz-rich sandstones and resulted from widespread chemical weathering of the Pan-African continental basement [4]. The composition of the sandstones changes from the arkose at the lower layers to sub-arkose and the younger layers are mainly mature quartz arenite [12, 13]. The fine fraction includes minerals that were formed by disintegration of the initial phases or result from dust storms.

Clay minerals usually cover the quartz grains. In the samples studied the dominant clay mineral is mainly kaolinite, small amounts of illite and some smectite. Kaolinite is usually crystallized at permeable bedrock in warm, moist regions forming as a residual weathering product. Illite might result from weathering of muscovite or K-feldspar [14], and smectite was formed from weathering of other minerals like biotite or amphibole that were in the Paleozoic sandstones from the first cycle. The crystal size of kaolinite is around 200-500 nm. Quartz-arenite stones from the Lower Miocene Moghra in Egypt were studied showing similar results with smectite, illite and kaolinite as clay minerals and hematite as the iron oxide [15].

Iron oxides were goethite, hematite, and some magnetite and maghemite. The iron oxides are responsible for the colors of the sandstones. Similar results were found in sands on the Atlantic coastal plain as iron oxides coat sand grains [16]. Goethite is usually yellow but impurities might change its color. Crystal's size might also affect the color. Goethite crystal sizes of 300 nm–1000 nm cause a yellow color; with smaller crystals it becomes darker. Hematite is usually yellow-red but larger crystals cause the appearance of a purple color. Formation of the iron oxides depends on the environment in which they were crystalized. Goethite usually crystallizes at fast oxidation at lower pH [17]. Hematite results from recrystallization of goethite close to a clay layer that adsorbs the OH of goethite, and hematite crystallizes (**Figure 2a**). Hematite and ilmenite ($FeTiO_3$) form a solid solution. Disintegration of ilmenite formed hematite crystals with Ti impurity and the leftover of the ilmenite was enriched with Ti forming $Fe/Ti < 1$ ratios.

Rutile (TiO_2) and anatase (TiO_2) had euhedral morphology, indicating that they might have crystallized in the area or, due to their hardness, they preserved their initial morphology. Rutile was found in the Paleozoic sandstone along with tourmaline [18]. Fe impurity in Ti-oxides might indicate that they were formed by in the area as ilmenite disintegrated and they were preserved within the surrounding clay minerals. It is also possible that rutile and anatase recrystallized from biotite from the magmatic and metamorphic rocks from the basement [19].

Jarosite ($KFe_3(SO_4)_2(OH)_6$) and alunite ($KAl_3(SO_4)_2(OH)_6$) were observed in dark violet or dark red sandstones. Jarosite was found in younger marl layers of Taqiya formation in the southern part of Israel, as a result of alteration of pyrite [20]. It was also found in Jurassic sandstone in the Ramon crater, and two possible formation processes were suggested: acid rain or from transgression, for a short period, of the sea [21]. Usually jarosite and alunite form a solid solution and they crystalize in saline lakes or acid sulphate soils [22]. In Utah (USA) they form cement in Jurassic sandstones and they precipitated in marginal marine to coastal dune [23]. Jarosite usually formed at low pH conditions and it requires an arid environment to prevent its decomposition into ferric oxyhydroxides [24]. It is possible that both jarosite and alunite were crystallized due to short transgression episodes of the Thetis Sea.

5. Conclusions

Quartz arenite sandstones were formed during Lower Cretaceous in the southern and northern parts of Israel. The fine fraction observed by TEM includes clay minerals mainly kaolinite and small amounts of illite and smectite. Additional minerals that contribute to the colors of the sandstones are iron oxides, goethite and hematite. The initial phase of iron oxides that coat quartz grains or form clusters was goethite, and hematite crystals preserved the initial acicular goethite structure. Ilmenite preserved within clay minerals disintegrate into small crystals enriched with iron. The tiny euhedral crystals of Ti oxides were probably formed in the sand. Jarosite and alunite observed in Ramon and Hatira craters were crystallized in the sandstones due to short transgression of the Thetis Sea.

Acknowledgements

The project was sponsored by Ilse Katz Institute for Nanoscale Science and Technology, Ben-Gurion University of the Negev, Beer-Sheva, Israel 8410501.

Author details

Nurit Taitel-Goldman[1*] and Vladimir Ezersky[2]

1 The Open University, Raanana, Israel

2 Ilse Katz Institute for Nanoscale Science and Technology, Ben-Gurion University of the Negev, Beer-Sheva, Israel

*Address all correspondence to: nurittg@hotmail.com

IntechOpen

© 2021 The Author(s). Licensee IntechOpen. This chapter is distributed under the terms of the Creative Commons Attribution License (http://creativecommons.org/licenses/by/3.0), which permits unrestricted use, distribution, and reproduction in any medium, provided the original work is properly cited.

References

[1] Bentor Y.K. and Vroman A., 1960, The geological map of Israel on a 1:100000 scale. Series A – The NegeV Sheet 16: mount Sdom (with explanary text) *Geol. Surv. Isr.*, 117p.

[2] Frank R. and Benjamini C., 2017, Field excursion in the Hatira anticline, northern Negev: High resolution sequence stratigraphy of the Aptian-Albian Hatira Succession. *Field trip guides Israel geological society, Mitzpe Ramon*, e1-e33.

[3] Lewy Z., 2003, Lower Aptian trace fossils in marginally marine sandstone and their extant analogues. *Isr. J. Earth Scie*. 52 39-46.

[4] Avigad D., Sandler A., Kolodner K., Stern R.J., McWilliamsM., Miller N. and Beyth. M., 2005, Mass-production of Cambro–Ordovician quartz-rich sandstone as a consequence of chemical weathering of Pan-African terranes: Environmental implications. *Earth and Planetary Science Letters.* 240 818-826.

[5] Kolodner K., Avigad D., Russel Ireland T. and Garfunkel Z., 2009, Origin of Lower Cretaceous ('Nubian') sandstones of North-east Africa and Arabia from detrital zircon U-Pb SHRIMP dating. Sedimentology 56 2010-2013.

[6] Wiessbrod T., 2002, stratigraphy and correlation of the lower Cretaceous exposures across the Dead Sea Transform with emphasis on tracing the Amir formation in Jordan. *Isr. J. Earth Sci.* 51 55-78.

[7] Tari G. et al. (2020) The Lower Cretaceous Chouf Sandstone of Lebanon: A Regional Reservoir Level in the Levant?. In: Khomsi S., Roure F., Al Garni M., Amin A. (eds) Arabian Plate and Surroundings: Geology, Sedimentary Basins and Georesources. Regional Geology Reviews. Springer, Cham. https://doi-org.elib.openu.ac.il/10.1007/978-3-030-21874-4_2

[8] Zilberman E., 2000, Formation of "makhteshim" - Unique erosion cirques in the Negev, southern Israel. Isr. J. Earth Sci. 49 3 127-141.

[9] Schindler M. and Singer D.M., 2017, Mineral surface coatings: environmental records at the nanoscale. Elements 13 159-164.

[10] Taitel-Goldman N., 2013, Recrystallization processes involving iron oxides in natural environments and In vitro, in: P. Wilson (editor) *Recent developments in the study of recrystallization.* InTech - open science - open minds, Rijeka, Croatia ISBN 978-953-51-0962-4

[11] Singer A., 1975, A Cretaceous laterite in the Negev Desert, Southern Israel. Geological magazine 112 151-162.

[12] Weissbrod, T., 1980, The Paleozoic of Israel and adjacent countries (a lithostratigraphic study): *(Ph.D. thesis) Jerusalem, Israel, Hebrew University of Jerusalem*, 255pp.

[13] Weissbrod, T. Perath I., 1990, Criteria for the recognition and correlation of sandstone units in the Precambrian and Paleozoic-Mesozoic clastic sequence in the Near East. *J. Afr. Earth Sci.* 10 253-270.

[14] Shelukhina O., El-Ghali M., AbbasiI.A., Hersi O.S. Farfour M., Ali A., Al-AwahH., Siddiqui N. A., 2021, Origin and control of grain-coating clays on the development of quartz overgrowths: example from the lower Paleozoic Barik Formation sandstones, Huqf region, Oman, 2021, Arab J Geosci. 14, 210.

[15] Tawfik H.A., Salah M.K. , Maejima W., Armstrong-Altrin J.S.,

Abdel-Monem T., A.H., El Ghandour M.M., 2018, Petrography and geochemistry of the Lower Miocene Moghra sandstones, Qattara Depression, north Western Desert, Egypt, *Geological journal* 53 1938-1953.

[16] Penn RL., Zhu C., Xu H., Veblen DR., 2001, Iron oxide coatings on sand grains from the Atlantic coastal plain. High resolution Transmission electron microscopy characterization. Geology 29 843-846.

[17] Cornell, R.M. and Schwertmann U., 2003, The Iron oxides, structure, properties, reactions, occurrences and uses. *Wiley-VCH verlagGmbH&Co. KGaA, Wieinheim.*

[18] Weissbrod, T., Nachmias J., 1987, Stratigraphic significance of heavy minerals in the Late Precambrian-Mesozoic clastic sequence (Nubian Sandstone) in the near east. Sediment. Geol. 47 263-291.

[19] Morad S. and Aldahan A., 1982, Authogenesis of titanium minerals in two Proterozoic sedimentary rocks from southern and central Sweden. *Journal of sedimentary petrology* 52. 4, 1295-1305.

[20] Sass E., Nathan Y., Nissenbaum A., 1965, Mineralogy of certain pyrite concretions from Israel and their alteration products. *Mineralog.* Mag, 35 84-87.

[21] Goldbery R., 1978, Early diagenetic nonhydrothermal Na-alunite in Jurassic flint clays, Makhtesh Ramon. Israel. Geol. Soc.Am. bull. 89 687-698.

[22] Jones F., 2017, Crystallization of Jarosite with variable Al^{3+} content: the transition to Alunite. MDPI Minerals 7 90.

[23] Potter-McIntyre S.L. and McCollom T.M., 2018, Jarosite and Alunite in ancient terrestrial sedimentary rocks: reinterpreting Martian depositional and diagenetic environmental conditions. MDPI Life 8 32.

[24] Bell J. H., Beitler Bowen, Martini, 2010, Imaging spectroscopy of jarosite cement in the Jurassic Navajo Sandstone. Remote Sensing of Environment 114 2259-2270.

Chapter 7

Diluted Magnetic Semiconductors Nanocrystals: Saturation and Modulation

Anielle C.A. Silva, Amanda I.S. Barbosa, Alessandra S. Silva, Elisson A. Batista, Thaís K. de Lima Rezende, Éder V. Guimarães, Ricardo S. Silva and Noelio O. Dantas

Abstract

Diluted Magnetic Semiconductor (DMS) nanocrystals are a new class of materials formed by doping the semiconductor with transition metals (TM), which gives interesting magneto-optical properties. These properties are attributed to the exchange interaction between the pure semiconductor's sp-electrons and the localized TM d-electrons. This book chapter shows exciting results of new DMS developed by the group, both in powder form and embedded in glassy systems. Depending on the concentration of doping ions, saturation of the incorporation of substitutional and interstitial sites in the nanocrystal structure may occur, forming other nanocrystals. In this context, we investigated the doping saturation limit in nanopowders of DMS $Zn_{1-x}Mn_xO$ NCs and $Zn_{1-x}Mn_xTe$, $Zn_{0.99-x}Mn_{0.01}Co_xTe$, and $Bi_{2-x}Co_xS$ NCs synthesized in glassy matrices. Thus, the sites' saturation into the crystalline lattice of nanocrystals is a topic little reported in the literature, and we will comment on this work. Therefore, we will show results from the group about the modulation and saturation in diluted magnetic semiconductors nanocrystals in this work.

Keywords: Diluted magnetic semiconductor, Nanocrystals, Saturation, Modulation, Doping, Transition metal

1. Introduction

The doping of semiconductors with small concentrations of transition metals (TM) ions, such as Co^{2+}, Cr^{3+}, Fe^{3+}, Mn^{2+}, allows the formation of new types of materials that have interesting spin-dependent electrical, optical, magnetic, and structural properties [1–3]. These new properties are attributed to the sp-d exchange interactions that involve the d-sub-levels of transition metal (TM) ions and the sp-electrons of the conduction band and/or holes in the host semiconductor valence band [4, 5]. These materials are called diluted magnetic semiconductors (DMS) and present great possibilities for technological applications such as the production of light-emitting diodes (LEDs) [6], spin transistors [7], lasers [8], supercapacitor [9], among others.

DMS materials, when developed under the quantum confinement regime, form nanocrystals (NCs) with smaller dimensions than the bulk material [10]. These DMS NCs have chemistry and physical properties dependent on their shape and size. These materials are obtained from a controlled process known as thermal diffusion of precursor ions to form DMS NC under the requirement of thermodynamic equilibrium.

The substitutional and interstitial sites' saturation in the nanocrystals structure may occur depending on the dopant concentration, and other nanocrystals types are formed. This system is called a nanocomposite and can have several exciting physical, chemical, and biological properties [11].

In this context, we investigated the doping saturation limit in nanopowders of DMS $Zn_{1-x}Mn_xO$ NCs [12] and $Zn_{1-x}Mn_xTe$ [13], $Zn_{0.99-x}Mn_{0.01}Co_xTe$ [14], and $Bi_{2-x}Co_xS$ [15] NCs synthesized in glassy matrices by the fusion method. The properties of the nanomaterials were investigated by experimental techniques of photoluminescence (PL), UV–Vis spectroscopy, X-ray diffraction (XRD), scanning electron microscopy (SEM), and transmission electron microscopy (MET). The theoretical study applying the crystalline field theory and Uv–Vis spectroscopy data allows identifying the tetrahedral (Th) or octahedral (Oh) location that the TM ions occupy in the crystalline structure of the semiconductor nanocrystal.

2. Diluted magnetic semiconductors in Nanopowders and embedded in glassy matrices

2.1 Synthesis of Nanopowders and nanocrystals and embedded in glassy matrices

ZnO NCs were synthesized by precipitation method using zinc nitrate and sodium hydroxide (NaOH, ≥98%) as precursors. In this work, the aqueous solution (1 M) of zinc nitrate (Zn(NO3)2. 6H2O, 98%) and the solution (2 M) of NaOH were mixed at room temperature. The NaOH solution was slowly added into zinc nitrate solution under vigorous stirring, which resulted in the formation of a white suspension. The white product was centrifuged at 6000 rpm for 5 min and washed several times with distilled water until the pH of the solution is around 7. The obtained samples were dried at 100°C for 24 hours. The samples were not subjected to temperatures above 100oC to avoid diffusion of Mn2+ ions from the nucleus to the surface of the ZnO NCs [15], since this is not the focus of this work. The ZnO:x Mn NCs were synthesized using the same procedure as ZnO NCs, but with the manganese (II) chloride (MnCl2, 98%) solution during the synthesis process. The xMn- concentration was determined based on the mass percentage of Zn present in ZnO (wt%), for x = 0.1; 0.3; 0.5; 0.7; 0.9; 1.0; 3.0; 5.0; 7.0 and 9.0. All reagents are nearly pure and purchased from Sigma-Aldrich Company.

$Zn_{1-x}Mn_xTe$ and $Zn_{0.99-x}Mn_{0.01}Co_xTe$ were synthesized by fusion method in a glass matrix and annealed post-growth. The fusion method consists of two sequential melting-nucleation processes that produce ensembles of nearly spherical nanoparticles embedded in a glass matrix. In the first step, the PZABP glass matrix with a nominal composition of $65P_2O_5 \cdot 14ZnO \cdot 1Al_2O_3 \cdot 10BaO \cdot 10PbO$ (mol %) adding 2Te (wt %), and Mn and/or Co at doping x content varying with Zn content from 0 to 80 (wt %), were synthesized by fusion in alumina crucibles at 1300°C for 30 minutes. Next, these melted mixtures were quickly cooled to room temperature forming a glass system doped with the precursor ions needed for nanoparticle growth. In the second step, the glass samples were thermally annealed at 500°C for 10 hours to enhance the diffusion of Zn^{2+}, Mn^{2+} and/or Co^{2+} and Te^{2-} ions throughout the host PZABP matrix and induce the growth of $Zn_{1-x}Mn_xTe/Zn_{0.99-x}Mn_{0.01}Co_xTe$ NCs. The physical properties of the glass samples were studied by optical absorption (OA), recorded with a model UV-3600 Shimadzu UV–VIS–NIR spectrometer, operating between 190 and 3300 nm; Photoluminescence (PL), using a 405 nm (~ 3.06 eV) continuous wave laser; Transmission electron micrographs (TEM JOEL, JEM-2100, 200 kV) and electron paramagnetic resonance (EPR), using a high sensitivity Bruker ESP-300 spectrometer (operating at X-band ~9.75 GHz and at Q-band ~34GHz).

Co^{2+}-doped Bi_2S_3 NCs were synthesized in a glass matrix with nominal composition of $45SiO_2$. $30Na_2CO_3$.$1Al_2O_3$. $24B_2O_3$ (mol%) adding 2S (wt%), 2Bi (wt%) and x = 0, 1 and 5% of Cobalt (Co), as a function of Bismuth (Bi) concentration. Samples were produced with the powder mixture added in an alumina crucible, and placed in a furnace at 1200 C for 30 min, followed by a fast cooling of the melted mixture down to room temperature, which formed a host glass. Thermal annealing was the next step: The previously-melted glass matrix, as-growth, was then heated at 500 C for 2, 10 and 24 h to enhance the diffusion of Bi^{3+} Co^{2+}, and S^{2-} ions precursors within the hosting matrix, resulting in the formation of DMS $Bi_{2-x}Co_xS_3$ NCs.

2.2 A study in function of Mn concentration in ZnO nanocrystals: Solubility and Mn^{2+} ions localization

The zinc oxide (ZnO) is a semiconductor with several interesting physical and biological properties such as ultraviolet absorption and bactericidal and antitumor properties [16]. Depending on the ions incorporated in the crystalline structure of ZnO nanocrystals (NCs) one can intensify or generate new and interesting properties. Dantas et al. demonstrated that doped ZnO nanocrystals' bactericidal and antitumor properties are potentiated or inhibited with the concentration and type of dopant [17]. The ZnO is a semiconductor of family II-VI that has a bandgap of 3.44 eV [18]. Due to its wide bandgap range, ZnO is suitable for technological applications of photonic devices operating in the blue and ultraviolet region, fabrication of nanodevices electronics [16, 19].

The structural and optical properties are shown in **Figure 1**. The characteristic diffraction patterns of ZnO crystals with wurtzite structure (JCPDS n° 36–1451) were observed in the X-ray patterns of samples (**Figure 1a**). The presence of additional diffraction peaks for concentrations above 3.0 Mn (%wt) corresponding to the phase $ZnMn_2O_4$ (JCPDS n° 24–1133), indicating the formation of the second phase (open circles). The closed circles are identified by the aluminum sample holder. Thus, this new crystal formation is associated with the saturation of the substitutional and interstitial sites in the ZnO crystal structure.

In order to identify the symmetrical environment in which the Mn^{2+} ions may be found in the ZnO NCs, adjustments will be made to the optical absorption spectra of the samples (ZnO, 0.5Mn, 1.0Mn and 9Mn) shown in **Figure 1(b)**. These adjustments are made in the positions of absorption of Mn^{2+} ions, making possible the use of the crystalline field theory (CFT) based on the crystalline field strength parameters Δ and Racah B, both calculated with help of Tanabe– Sugano diagram (**Figure 1(c)**) [20]. From the OA spectra of the NCs of Zn: xMn, based on the Tanabe-Sugano diagram, the energies of the characteristic electronic transitions of Mn^{2+}: ${}^6A_1({}^6S) \rightarrow {}^4T_1({}^4G)$ (691 nm), ${}^6A_1({}^6S) \rightarrow {}^4T_2({}^4G)$ (590 nm), ${}^6A_1({}^6S) \rightarrow {}^4T_2({}^4D)$ (482 nm), and ${}^6A_1({}^6S) \rightarrow {}^2E({}^2I)$ (410 nm) (**Figure 1b**) subtly permitted by spin-orbit coupling were effectively described by the Racah B parameter (559 cm^{-1}) and the crystalline field division (Δ = 5464 cm^{-1}) [21, 22].

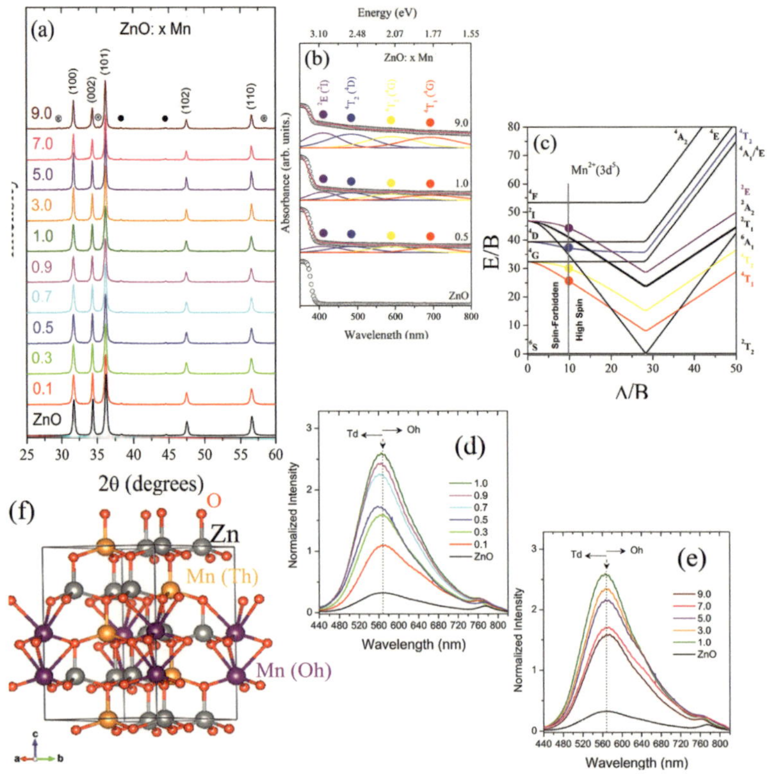

Figure 1.

(a) X-ray diffractograms (b) As adjusted curves represent the experimental energies of the transitions prohibited by spin $^6A_1$ *(S)* $\rightarrow$ $^4T_1$ *(G),* $^4T_2$ *(G),* $^4T_2$ *(D),* $^2A_2$ *(I) e* ^2E*(I). (c) Tanabe-Sugano diagram for the electronic configuration of the Mn2+ (3d5) (C/B = 4.5) with vertical solid line in* $\Delta/B = 9.77$*,* $\Delta = 5464\ cm^{-1}$*, and* $B = 559\ cm^{-1}$*, PL spectra from 440 to 820 nm for concentrations of (d) 0.1 to 1.0 of Mn and (e) of 1.0 to 9.0 of Mn, (f) wurtzite structure of ZnO NCs with coordination geometry in which the* Mn^{2+} *ions are, represented by tetrahedral sites (td) and octahedral sites (od).*

The results from Δ and B show that the Mn^{2+} dopants incorporated mainly at octahedral sites in the hexagonal structure of ZnO, for the sample with low concentration, and in the tetragonal structure of $ZnMn_2O_4$. Thus, the excited states of the Mn^{2+} in the binder field reside in the host semiconductor energy gap [21, 22].

Fluorescence spectra of the nanocrystals of ZnO: x Mn where the band at 570 nm are assigned to vacancies [23] are shown in **Figure 1(d, e)**. Luminescence intensification and *blue shift* to a concentration of 0.5 Mn. This result confirms the incorporation of Mn^{+2} ions into the tetrahedral sites of ZnO NCs (inside nanocrystal). However, there is a contribution to incorporating Mn^{2+} ions into the octahedral sites (surface nanocrystal). It is noteworthy that the band widening around 570 nm is due an overlapping characteristics bands of oxygen vacancies and d–d ($^4T_1 \leftarrow {}^6A_1$) of Mn^{2+}-ions in ZnO NC host [23, 24].

As the concentration increases from 0.5 to 1.0 Mn, the band's position remains, confirmed the half-height with x_{Mn}-concentration, which is related to the presence of the formation of the $ZnMn_2O_4$ NCs and that the Mn^{2+} are incorporated in the octahedral sites of the crystal structure saturation of the sites present in the ZnO NCs.

In order to visually check how these are coordination geometry, **Figure 1(f)** shows the wurtzite structure of ZnO NCs, which the Mn^{2+} ions are in tetrahedral sites (Td) and/or octahedral sites (Od).

2.3 Solubility of manganese and cobalt in ZnTe nanocrystals embedded in P_2O_5 - ZnO - Al_2O_3 - BaO - PbO glassy matrix

The incorporation of magnetic ions into the crystal structure of ZnTe semiconductor nanocrystals, which grow in glassy systems, has become very interesting due to the various applications in spintronics are governed by sp–d spin interaction between carriers and magnetic ions [25]. Some possible applications require nanoparticles to be embedded in highly stable, robust, and transparent host materials, as glassy systems [26]. The present study reports on the investigation of the solubility/saturation of Mn^{2+} ions in $Zn_{1-x}Mn_xTe$ [13, 26, 27] and competition between Co^{2+} and Mn^{2+} ions into $Zn_{0.99-x}Mn_{0.01}Co_xTe$ [14, 28] nanosized DMS embedded in a glassy system with a wider Mn/Co concentration range (x = 0.000–0.800). According to the literature, the Mn^{2+}/Co^{2+} ion can be thermodynamically incorporated into II-VI semiconductors up to its solubility limit [29, 30]. Above this limit, saturation occurs, and Mn^{2+}/Co^{2+} ions tend to be expelled toward the semiconductor nanocrystal surface [30].

Figure 2 presents OA (a) and PL (c) spectra and photographs (b) of the samples containing $Zn_{1-x}Mn_xTe$ NCs, with Mn-concentrations ranging from $x = 0.000$ to $x = 0.800$, and with PZABP ($65P_2O_5 \cdot 14ZnO \cdot 1Al_2O_3 \cdot 10BaO \cdot 10PbO$ (mol %)) glassy matrix doped only 80Mn (% wt of Zn). The OA spectra show absorption bands centered attributed to $Zn_{1-x}Mn_xTe$ NCs (quantum dots (QDs) and bulk NCs) and the incorporation of Mn^{2+} ions. This substitutional incorporation is confirmed from the EPR spectra (**Figure 2(f)**) that showed six lines associated to S = 5/2 spin half-filled d-state, characteristic of Mn^{2+} ions (as detailed in the energy diagram) [26]. For x > 0, additional bands are observed.

The absorption band centered at 3.54 eV is attributed at MnO_2, and the bands centered at 3.02, and 2.43 eV is attributed at MnO [31]. The TEM images in **Figure 2(e)** (for x = 0.100), revealed interplanar distances corresponding to $Zn_{0.9}Mn_{0.1}Te$ NCs (d ~ 0.347 nm) [32], MnO (d ~ 0.224 nm) [33] and MnO_2 (d ~ 0.582 nm) NCs [34]. This result can be confirmed by the PL spectra, on what the observed redshift shows that it was possible to tune Mn emission energy from orange to near-infrared as a function of concentration. These results confirm the successful inclusion of Mn^{2+} ions in the $Zn_{1-x}Mn_xTe$ up to the nominal solubility limit of x = 0.100. Above this solubility limit, one can observe Mn's saturation, forming MnO and MnO_2 NCs, as represented in **Figure 2(d)**. Thus, one of the main motivations for studying semiconductors doped with transition metals, especially Mn^{2+} ions, is for applications in luminescent devices.

Figure 3 presents OA (a), PL (b) and EPR (c) spectra and photographs (between OA and PL spectra) of the samples containing $Zn_{0.99-x}Mn_{0.01}Co_xTe$ NCs, with Co-concentrations ranging from $x = 0.000$ to $x = 0.800$. **Figure 3(d)** presents the energy diagram of all transitions related to OA and PL data, and **Figure 3(e)** shows the transitions satisfying the selection rules $\Delta M_S = \pm 1$ and $\Delta M_I = 0$. The AO spectra, besides the bands attributed to $Zn_{1-x}Mn_xTe$ QDs and bulk NCs, as previously presented in **Figure 3**, four characteristic bands in the visible spectrum for all doped samples with cobalt. These four are due to the spin-allowed transitions: ${}^4A_2(F) \rightarrow {}^2T_1(P)$ (2.13 eV); and spin-forbidden transitions: ${}^4A_2(F) \rightarrow {}^2A_1(G)$ (2.33 eV), ${}^4A_2(F) \rightarrow {}^4T_1(G)$ (1.95 eV) and ${}^4A_2(F) \rightarrow {}^2E(G)$ (1.91 eV) [14]. These bands show the substitutional incorporation of Co^{2+} ions in tetrahedral Zn^{2+} ions sites. The PL spectra show emissions from two NC groups: QDs (E_{exc} ~ 2.53 eV) and bulk NCs (E_b ~ 2.21 eV); defects related to zinc vacancies (E_{VZn} ~ 2.53 eV) and oxygen centers (E_o ~ 1.91 eV) [14, 27].

The emission centered at approximately 2.03 eV and 1.89 eV corresponds to tetrahedrally (Td) and octahedrally (Oh) coordinated Mn^{2+} ions. Already, when the

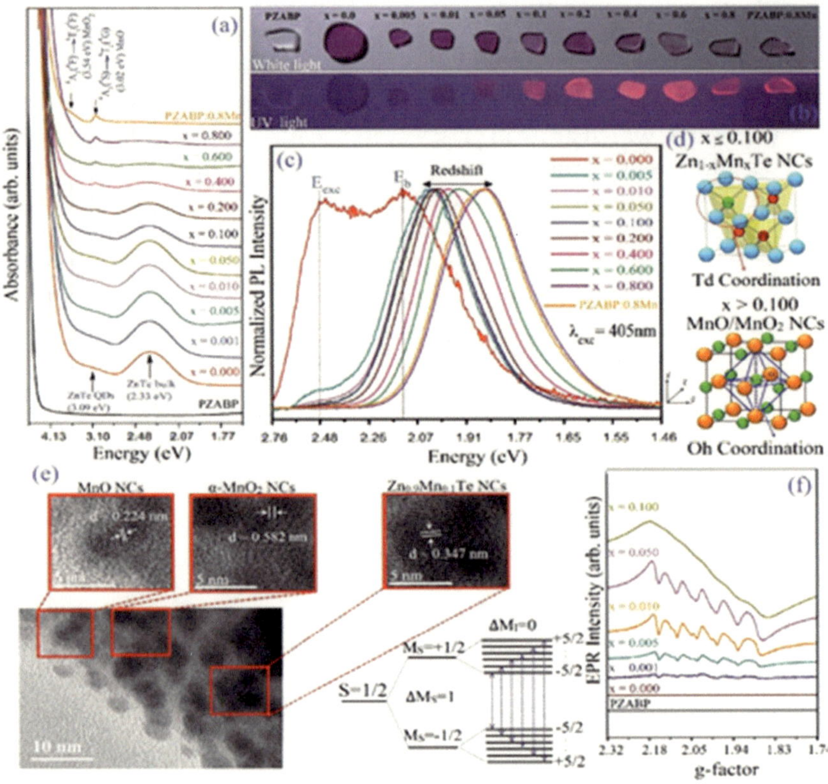

Figure 2.
OA (a) and PL (c) spectra and photographs (b) of samples containing $Zn_{1-x}Mn_xTe$ NCs, with Mn-concentrations ranging from x = 0.000 to x = 0.800. In (d) Mn^{2+} ions located in tetrahedral sites (td) of ZnTe NCs (concentration x < 0.100) and Mn^{2+} and Mn^{4+} located in octahedral sites (oh) of MnO and MnO_2 NCs (concentration x ≥ 0.100); (e) TEM images for concentration x = 0.100; (f) EPR spectra of $Zn_{1-x}Mn_xTe$ NCs with Mn concentrations varying from x = 0.00 to x = 0.100 and selection rules $\Delta M_S = \pm 1$ and $\Delta M_I = 0$ (next to).

spin-allowed transitions centred around 1.69 eV ($^4T_1(^4P) \rightarrow {}^4A_2(^4F)$, named E_{1Co2+}) and 1.52 eV ($^4T_1(^4P) \rightarrow {}^4T_2(^4F)$, named E_{2Co2+}) [14]. The energy diagram in **Figure 3(d)**shows these OA and PL spectra transitions. The EPR spectra, besides proving the substitutional incorporation of Mn^{2+} in $Zn_{1-x}Mn_xTe$, as justified by **Figure 3(f)**, also reveals that as the Co doping, an intense central line with g = 2.012 attributed to interacting Co^{2+} [14]. For both Mn^{2+} and Co^{2+} ions the transitions observed satisfy the selection rules $\Delta M_S = \pm 1$ and $\Delta M_I = 0$ (**Figure 3(e)**). Thus, the EPR spectra suggest a kind of competition between Mn^{2+} and Co^{2+} ions to substitute Zn^{2+} ions, controlled by the Co concentration.

2.4 Effects of symmetry and concentration of cobalt-doped Bi_2S_3 nanocrystals embedded in host glass matrix

Bi_2S_3 semiconductor nanocrystals (NCs) present radioactive recombination of electron–hole carriers adjustable according to the temperature, time, and concentration of doping ions in the system [15, 35–37]. The occupation of vacancies in the Bi_2S_3 orthorhombic network by Co^{2+} ions (0.72 Å) replacing Bi^{3+} ions (1.03 Å) can reduce intrinsic point defects and consequently increase the efficiency of semiconductor nanocrystals, for applications in solar cells [38], photocatalytic devices of visible light [39], thermoelectric [40] and spintronic [41]. Therefore, the control of the magnetic saturation of the energy states located of the Co^{2+} ions in the Bi_2S_3 NCs

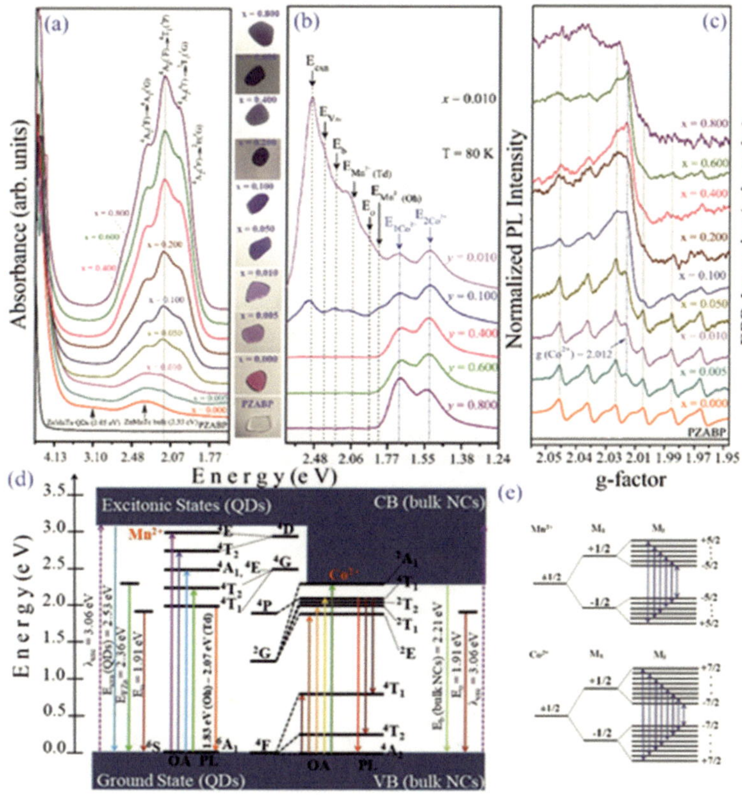

Figure 3.
OA (a), PL (b) and EPR (c) spectra and photographs (between OA and PL spectra) of the samples containing $Zn_{0.99-x}Mn_{0.01}Co_xTe$ NCs, with Co-concentrations ranging from x = 0.000 to x = 0.800. Energy diagram of all transitions related to OA and PL data (d). Transitions satisfying the Mn^{2+} and Co^{2+} selection rules $\Delta M_S = \pm 1$ and $\Delta M_I = 0$ (e).

bandgap allows assigning new magnetic and optoelectronic properties. In this context, we investigated the effect of the molar fraction and the coordination symmetry of Co^{2+} ions in cobalt-doped Bi_2S_3 NCs embedded in a host glass matrix (SNAB - $45SiO_2 \cdot 30Na_2CO_3 \cdot 5Al_2O_3 \cdot 20B_2O_3$) (mol%).

Figure 4(a) shows the UV–VIS–NIR optical absorption spectrum (OA) of the $Bi_{2-x}Co_xS_3$ NCs embedded in a SNAB glass matrix, at temperature (300 K), as a function of the increasing concentration of doping xCo ions (x = 0.000; 0.005; 0.010; 0.050; 0.100). The OA spectrum of the SNAB matrix is transparent in the spectral region where the $Bi_{2-x}Co_xS_3$ NCs can absorb and emit a photon. The OA spectrum of the samples shows energy bands of the exciton charge carrying states and the d-d transitions of the Co^{2+} ions in the $Bi_{2-x}Co_xS_3$ NCs. The structure of the absorption bands is due to that of the ligand field of S^{2-} ions in distorted tetrahedral sites around the Co^{2+} ions. Such bands of electronic level transitions have increased due to the increasing concentration of Co^{2+} ions in the Bi_2S_3 NCs structure.

Furthermore, the p-d and spin-orbit interactions in an asymmetric tetrahedral coordination complex $[CoS_4]^{6-}$ provide a higher intensity for 3d-3d transitions [14, 15, 20, 42]. Therefore, the exotic properties of these new materials called diluted magnetic semiconductors (DMS) NCs are linked to the nature of the sp-d exchange interaction that occurs between the charge carriers of the Bi_2S_3 semiconductor bands around the electrons in the central metal ion orbital (Co^{2+}) with $3d^7$ electronic configuration (spin) $3d^7$ (S = 3/2) [15, 20, 36, 42].

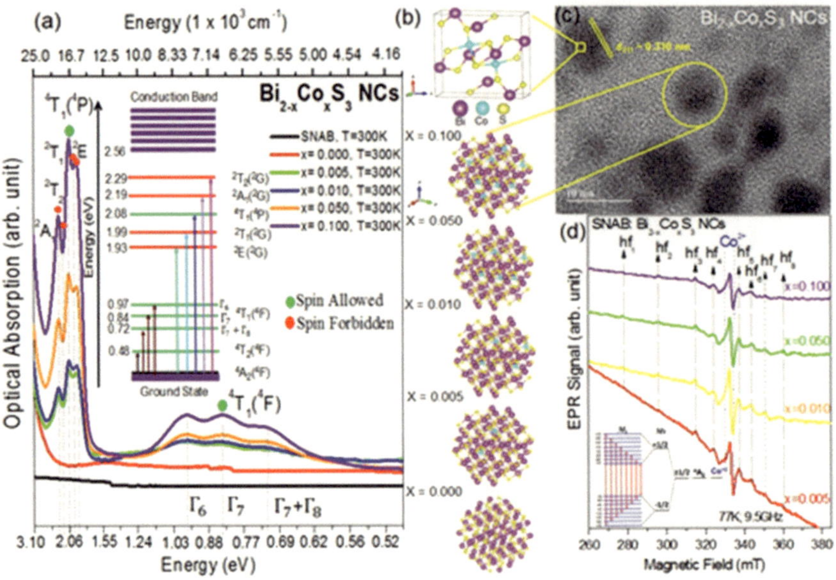

Figure 4.
(a) Optical absorption spectra at room temperature of $Bi_{2-x}Co_xS_3$ NCs (x = 0.000; 0.005; 0.010; 0.050; 0.100) embedded in the SNAB glass matrix annealed for 2 h at 500°C. for comparison purposes, the absorption spectrum of the SNAB glass matrix is represented on the black bottom line. The inset shows the energy level diagram for Co^{2+} ($3d^7$) in a tetrahedral site and the respective spin allowed and forbidden transitions. (b) Bi_2S_3 orthorhombic unit cell and $Bi_{2-x}Co_xS_3$ quantum dots with the interstitial replacement of Bi^{3+} ions by Co^{2+} at distorted tetrahedral sites. (c) TEM image of $Bi_{2-x}Co_xS_3$ NCs (x = 0.100). (d) EPR spectra (77 K) of $Bi_{2-x}Co_xS_3$ NCs (x = 0.005; 0.010; 0.050; 0.100) embedded in the SNAB glass matrix.

For $Bi_{2-x}Co_xS_3$ NCs, the spin allowed transitions $^4A_2$ (^4F) → $^4T_1$ (^4F) and $^4A_2$ (^4F) → $^4T_1$ (^4P) are obtained with the energy value of the barycenter for a normal distribution of the bands observed in the spectrum absorption in the range of 0.62–1.24 eV and 1.70–2.60 eV. The bands observed in the OA spectrum of **Figure 4(a)** correspond to the spin allowed and forbidden transitions, $^4A_2$ (^4F) → $^4T_1$ (^4F) (0.84 eV), $^4A_2$(^4F) → ^2E (^2G) (1.93 eV), $^4A_2$(^4F) → $^2T_1$(^2G) (1.99 eV), $^4A_2$(^4F) → $^4T_1$(^4P) (2.08 e V), $^4A_2$(^4F) → $^2A_1$(^2G) (2.19 eV) and $^4A_2$(^4F) → $^2T_2$(^2G) (2.29 eV). Such identified 3d-3d transitions are attributed to the crystal field strength (Δ = 3882 cm^{-1}) and the electronic repulsion parameter (Racah, B = 772 cm^{-1}), based on the crystal field theory and the Tanabe-Sugano diagram d^7, tetrahedral for C/B = 4.5 [20, 43, 44]. The wide absorption band in the infrared is due to a spin-orbit coupling interaction that split the excited state $^4T_1$ (^4F) into three energy sub-states: Ä$_6$ (0.97 eV), Ä$_7$ (0.84 eV) e Ä$_7$ + Ä$_8$ (0.72 eV) [12, 13]. Therefore, the excited states of the Co^{2+} ions are located in the energy gap (2.65 eV) of electrons and holes confined between the ground state and the conduction band of the $Bi_{2-x}Co_xS_3$ NCs. The energy level diagram of Co^{2+} ions (3d^7) in a ligand field of S^{2-} ions at tetrahedral coordination sites $[CoS_4]^{6-}$ is shown in the inset in Figure x (a). The energy band associated with the spin allowed transition $^4A_2$ (^4F) → $^4T_1$ (^4F) (0.48 eV) does not appear in the OA spectrum (**Figure 4a**) due to the low intensity and energy [15, 43].

The redshift observed at the OA band edge for the $Bi_{2-x}Co_xS_3$ NCs with increasing concentration xCo, from x = 0.000 (2.75 eV) to x = 0.100 (2.56 eV) (see **Figure 4a**), are consequences of sp-d exchange interactions between sp. electrons confined in states of quantum dots and localized states partially filled with 3d electrons of the Co^{2+} ions [10]. No significant changes occur in the quantum size of the $Bi_{2-x}Co_xS_3$ NCs with the incorporation of Co. The TEM image (**Figure 4c**) shows an average size of 6 nm for $Bi_{2-x}Co_xS_3$ NCs (x = 0.100) with quantum dot properties. The interplanar distance (d_{211} = 0.310 nm) shows the crystal plane (211) characteristic of the orthorhombic structure of the mineral Bismuthinite. **Figure 4(b)** illustrates the Bi_2S_3 orthorhombic structure unit cell doped with Co^{2+} ions at distorted tetrahedral sites and the respective $Bi_{2-x}Co_xS_3$ quantum dots function of the xCo concentration.

Figure 4(d) shows the EPR spectra obtained in the best conditions at 77 K and 9.75 GHz for samples of $Bi_{2-x}Co_xS_3$ NCs embedded in SNAB glass matrix, with xCo concentration ranging from x = 0.005 to 0.100. The intense central signal at ($g \approx 2.005$; $\Delta H \approx 8$ mT) corresponds to free Co^{+2} ions, dispersed in the glass matrix and not incorporated in the Bi_2S_3 NCs [2, 10]. The inset in Figure x (d) presents a diagram with the allowed hyperfine lines $\Delta M_S = \pm 1$ with $\Delta M_I = 0$, related to the transition between the levels of fine interaction $M_S = +½ \leftrightarrow -½$. The anisotropic characteristic of the eight lines (hf_1; hf_2; hf_3; hf_4; hf_5; hf_6; hf_7; hf_8) of the hyperfine interaction between the electron spin (S = 3/2) and the nuclear spin (I = 7/2) and its similar shapes for all xCo concentrations confirms magnetic doping in distorted tetrahedral sites. The result is compatible with a region where there is a large, inhomogeneous crystal field due to the lack of interface between the host glass matrix and the NCs [42].

3. Conclusion

The synthesis and investigation of the DMS NC have allowed the production of new types of materials with possible technological applications in Engineering, Medicine, Environment, Telecommunications, and others. We report the synthesis of the $Zn_{1-x}Mn_xO$ nanopowders and the Bi2-xCoxS3, Zn1-xMnxTe $Zn_{0.99-x}Mn_{0.01}Co_xTe$ NCs in glass by fusion method. The formation of DMS NCs is observed by SEM and MET images. The EPR spectra confirm the incorporation of TR^{2+} ions (Mn^{2+} and Co^{2+}) in the doped nanocrystals' crystalline structure. The optical properties of these materials were investigated by photoluminescence and Uv–Vis spectroscopy techniques, which show the influence of TR^{2+} ions (Mn^{2+} and Co^{2+}) on the visible spectrum, altering the optical absorption and photoluminescence bands of undoped naocrystals. In this work, we hope to contribute significantly to the study of DMS NCs and their possible technological applications for society's benefit.

Author details

Anielle C.A. Silva[1,2*], Amanda I.S. Barbosa[1], Alessandra S. Silva[5], Elisson A. Batista[3], Thaís K. de Lima Rezende[4], Éder V. Guimarães[5], Ricardo S. Silva[5] and Noelio O. Dantas[1]

1 Laboratory of New Nanostructured and Functional Materials, Physics Institute, Federal University of Alagoas, Maceió, AL, Brazil

2 Programa de Pós-Graduação da Rede Nordeste de Biotecnologia (RENORBIO), Federal University of Alagoas, Maceió, AL, Brazil

3 Physics Institute, Federal University of Uberlândia, Uberlândia, MG, Brazil

4 Chemical Institute, Federal University of Uberlândia, Uberlândia, MG, Brazil

5 Institute of Exact, Natural Sciences and Education (ICENE), Department of Physics, Federal University of Triângulo Mineiro, Uberaba, MG, Brazil

*Address all correspondence to: aniellechristineas@gmail.com

IntechOpen

© 2021 The Author(s). Licensee IntechOpen. This chapter is distributed under the terms of the Creative Commons Attribution License (http://creativecommons.org/licenses/by/3.0), which permits unrestricted use, distribution, and reproduction in any medium, provided the original work is properly cited.

References

[1] Lourenço, S.A.; Dantas, N.O.; Souza, R. On Synthesis, Optical and magnetic Properties of Semimagnetic Nanocrystals Growth in Glass Matrix. Nanomagnetism **2014**, 53-73.

[2] Silva, R.S. da; Neto, E.S. de F.; Dantas, N.O. Optical, Magnetic, and Structural Properties of Semiconductor and Semimagnetic Nanocrystals. In *Nanocrystals - Synthesis, Characterization and Applications*; 2012.

[3] Schulz, H.J. Transition metal impurities in compound semiconductors: Experimental situation - ScienceDirect. Mater. Chem. Phys. **1987**, *16*, 373-384.

[4] Oliveira, N.; Freitas Neto, E.S. De; Da Silv, R.S. Diluted Magnetic Semiconductor Nanocrystals in Glass Matrix. In *Nanocrystals*; 2010.

[5] Cibert, J.; Scalbert, D. Diluted Magnetic Semiconductors: Basic Physics and Optical Properties. In; Springer, Berlin, Heidelberg, 2008; pp. 389-431.

[6] Wang, C.; Yang, F.; Gao, Y. The highly-efficient light-emitting diodes based on transition metal dichalcogenides: From architecture to performance. Nanoscale Adv. 2020, *2*, 4323-4340.

[7] Rai, D.P.; Laref, A.; Shankar, A.; Sandeep; Sakhya, A.P.; Khenata, R.; Thapa, R.K. Spin-induced transition metal (TM) doped SnO2 a dilute magnetic semiconductor (DMS): A first principles study. J. Phys. Chem. Solids **2018**, *120*, 104-108, doi:10.1016/j.jpcs.2018.04.006.

[8] Morandi, O.; Hervieux, P.A.; Manfredi, G. Laser induced ultrafast demagnetization in diluted magnetic semiconductor nanostructures. In Proceedings of the European Physical Journal D; 2009; Vol. 52, pp. 155-158.

[9] Brozek, C.K.; Zhou, D.; Liu, H.; Li, X.; Kittilstved, K.R.; Gamelin, D.R. Soluble Supercapacitors: Large and Reversible Charge Storage in Colloidal Iron-Doped ZnO Nanocrystals. Nano Lett. **2018**, *18*, 3297-3302, doi:10.1021/acs.nanolett.8b01264.

[10] Dantas, N.O.; dos Silva, A.S.; Silva, A.C.A.; de Neto, E.S.F. Scanning probe microscopies of semiconductor and semimagnetic nanocrystals. In *Advances in Laser and Optics Research*; 2015 ISBN 9781634635226.

[11] Silva, A.; Zóia, M.A.P.; Correia, L.I.V.; Azevedo, F.V.P.V.; Paula, A.T. de; Maia, L.P.; Carvalho, L.S. de; Carvalho, L.N.; Costa, M.P.C.; Giaretta, L.C.; et al. Biocompatibility of Doped Semiconductors Nanocrystals and Nanocomposites. In *Cytotoxicity*; InTech, 2018.

[12] Batista, E.A.; Silva, A.C.A.; de Lima, T.K.; Guimarães, E.V.; da Silva, R.S.; Dantas, N.O. Effect of the location of Mn2+ ions in the optical and magnetic properties of ZnO nanocrystals. J. Alloys Compd. **2021**, *850*, 156611, doi:10.1016/j.jallcom.2020.156611.

[13] Silva, A. dos S.; da Silva, S.W.; de Morais, P.C.; Dantas, N.O. Solubility limit of Mn2+ ions in Zn1−xMnxTe nanocrystals grown within an ultraviolet-transparent glass template. *J. Nanoparticle Res.* 2016, *18*.

[14] Silva, A.S.; Lourenço, S.A.; Da Silva, M.A.T.; Da Silva, S.W.; Morais, P.C.; Dantas, N.O. Effect of Co co-doping on the optical properties of ZnTe:Mn nanocrystals. Phys. Chem. Chem. Phys. **2017**, *19*, 1158-1166, doi:10.1039/c6cp05866c.

[15] Guimarães, E. V.; Gonçalves, E.R.; Lourenço, S.A.; Oliveira, L.C.; Baffa, O.; Silva, A.C.A.; Dantas, N.O.; Silva, R.S.

Concentration effect on the optical and magnetic properties of Co2+-doped Bi2S3 semimagnetic nanocrystals growth in glass matrix. J. Alloys Compd. **2018**, *740*, 974-979, doi:10.1016/j.jallcom.2018.01.073.

[16] Borysiewicz, M.A. ZnO as a functional material, a review. Crystals **2019**, *9*, 505, doi:10.3390/cryst9100505.

[17] Silva, A.; Zóia, M.A.P.; Correia, L.I.V.; Azevedo, F.V.P.V.; Paula, A.T. de; Maia, L.P.; Carvalho, L.S. de; Carvalho, L.N.; Costa, M.P.C.; Giaretta, L.C.; et al. Biocompatibility of Doped Semiconductors Nanocrystals and Nanocomposites. In *Cytotoxicity*; InTech, 2018.

[18] Janotti, A.; Van De Walle, C.G. Fundamentals of zinc oxide as a semiconductor. Reports Prog. Phys. **2009**, *72*, 126501, doi:10.1088/0034-4885/72/12/126501.

[19] Mishra, Y.K.; Adelung, R. ZnO tetrapod materials for functional applications. Mater. Today **2018**, *21*, 631-651, doi:10.1016/j.mattod.2017.11.003.

[20] Lourenço, S.A.; Silva, R.S.; Silva, A.C.A.; Dantas, N.O. Structural and Optical Properties of Co2+-Doped PbSe Nanocrystals in Chalcogeneide Glass Matrix. J. Phys. Chem. C **2015**, *119*, 13277-13282, doi:10.1021/acs.jpcc.5b01920.

[21] Silva, A.S.; Lourenço, S.A.; Dantas, N.O. Mn concentration-dependent tuning of Mn2+ d emission of Zn1-xMnxTe nanocrystals grown in a glass system. Phys. Chem. Chem. Phys. **2016**, *18*, 6069-6076, doi:10.1039/c5cp06802a.

[22] Beaulac, R.; Archer, P.I.; Gamelin, D.R. Luminescence in colloidal Mn2+-doped semiconductor nanocrystals. J. Solid State Chem. **2008**, *181*, 1582– 1589, doi:10.1016/j.jssc.2008.05.001.

[23] Romeiro, F.C.; Marinho, J.Z.; A. Silva, A.C.; Cano, N.F.; Dantas, N.O.; Lima, R.C. Photoluminescence and Magnetism in Mn $^{2+}$ -Doped ZnO Nanostructures Grown Rapidly by the Microwave Hydrothermal Method. J. Phys. Chem. C **2013**, *117*, 26222-26227, doi:10.1021/jp408993y.

[24] Neto, E.S.F.; Dantas, N.O.; Neto, N.M.B.; Guedes, I.; Chen, F. Control of luminescence emitted by Cd1-xMnxS nanocrystals in a glass matrix: X concentration and thermal annealing. Nanotechnology **2011**, *22*, doi:10.1088/0957-4484/22/10/105709.

[25] Karan, N.S.; Sarma, D.D.; Kadam, R.M.; Pradhan, N. Doping transition metal (Mn or Cu) ions in semiconductor nanocrystals. J. Phys. Chem. Lett. **2010**, *1*, 2863-2866, doi:10.1021/jz1012164.

[26] Silva, A.S.; Franco, A.; Pelegrini, F.; Dantas, N.O. Paramagnetic behavior at room temperature of Zn1-xMnxTe nanocrystals grown in a phosphate glass matrix by the fusion method. J. Alloys Compd. 2015, *647*, 637-643.

[27] Silva, A.S.; Lourenço, S.A.; Dantas, N.O. Mn concentration-dependent tuning of Mn2+ d emission of Zn1-xMnxTe nanocrystals grown in a glass system. Phys. Chem. Chem. Phys. **2016**, *18*, 6069-6076, doi:10.1039/c5cp06802a.

[28] Silva, A.S.; Pelegrini, F.; Figueiredo, L.C.; De Souza, P.E.N.; Da Silva, S.W.; De Morais, P.C.; Dantas, N.O. Evidence of competition in the incorporation of Co2+ and Mn2+ ions into the structure of ZnTe nanocrystals. RSC Adv. **2016**, *6*, 101226-101234, doi:10.1039/c6ra19189d.

[29] Sayan Bhattacharyya; Zitoun, D.; Estrin, Y.; Moshe, O.; Rich, D.H.; Gedanken, A. A one-step, template-free synthesis, characterization, optical and magnetic properties of Zn1-xMnxTe nanosheets. *Chem. Mater.* **2009**, *21*, 326-335, doi:10.1021/cm8027737.

[30] Erwin, S.C.; Zu, L.; Haftel, M.I.; Efros, A.L.; Kennedy, T.A.; Norris, D.J. Doping semiconductor nanocrystals. Nature **2005**, *436*, 91-94, doi:10.1038/nature03832.

[31] Noginova, N.; Lindsay, W.; Noginov, M.A.; Loutts, G.B.; Mattix, L. Photorefractive effects in Mn-doped YAlO_3. J. Opt. Soc. Am. B **1999**, *16*, 754, doi:10.1364/josab.16.000754.

[32] Lin, Y.W.; Chen, W.J.; Lu, J.Y.; Chang, Y.H.; Liang, C. Te; Chen, Y.F.; Lu, J.Y. Growth and characterization of ZnO/ZnTe core/shell nanowire arrays on transparent conducting oxide glass substrates. Nanoscale Res. Lett. **2012**, *7*, 1-5, doi:10.1186/1556-276X-7-401.

[33] Liu, B.; Hu, X.; Xu, H.; Luo, W.; Sun, Y.; Huang, Y. Encapsulation of MnO Nanocrystals in Electrospun Carbon Nanofibers as High-Performance Anode Materials for Lithium-Ion Batteries. Sci. Rep. **2015**, *4*, 4229, doi:10.1038/srep04229.

[34] Wei, M.; Konishi, Y.; Zhou, H.; Sugihara, H.; Arakawa, H. Synthesis of single-crystal manganese dioxide nanowires by a soft chemical process. Nanotechnology **2005**, *16*, 245-249, doi:10.1088/0957-4484/16/2/011.

[35] Tarachand; Sharma, V.; Bhatt, R.; Ganesan, V.; Okram, G.S. A catalyst-free new polyol method synthesized hot-pressed Cu-doped Bi2S3 nanorods and their thermoelectric properties. *Nano Res.* **2016**, *9*, 3291-3304, doi:10.1007/s12274-016-1207-6.

[36] Guimarães, É. V.; Mikhail, H.D.; Silva, A.C.A.; Dantas, N.O.; Silva, R.S. Investigations of structural and optical properties of Bi2−xCrxS3 nanocrystals embedded in host glass. Mater. Lett. **2020**, *265*, doi:10.1016/j.matlet.2020.127430.

[37] da Silva, R.; Silva, J.; Rocha, V.; Cano, N.; Almeida Silva, A.; Dantas, N. Synthesis Process Controlled of Semimagnetic $Bi_{2-x}Mn_xS_3$ Nanocrystals in a Host Glass Matrix. *J. Phys. Chem. C* *118*, 18730-18735, doi:10.1021/jp5046657.

[38] Han, D.; Du, M.H.; Dai, C.M.; Sun, D.; Chen, S. Influence of defects and dopants on the photovoltaic performance of Bi2S3: First-principles insights. J. Mater. Chem. A **2017**, *5*, 6200-6210, doi:10.1039/c6ta10377d.

[39] Liu, Z.Q.; Huang, W.Y.; Zhang, Y.M.; Tong, Y.X. Facile hydrothermal synthesis of Bi 2S 3 spheres and CuS/Bi 2S 3 composites nanostructures with enhanced visible-light photocatalytic performances. CrystEngComm **2012**, *14*, 8261-8267, doi:10.1039/c2ce26123e.

[40] Biswas, K.; Zhao, L.D.; Kanatzidis, M.G. Tellurium-free thermoelectric: The anisotropic n -type semiconductor Bi 2S 3. Adv. Energy Mater. **2012**, *2*, 634-638, doi:10.1002/aenm.201100775.

[41] Zhao, G.; Deng, Z.; Jin, C. Advances in new generation diluted magnetic semiconductors with independent spin and charge doping. J. Semicond. 2019, *40*.

[42] Freitas Neto, E.S.; Silva, A.C.A.; da Silva, S.W.; Morais, P.C.; Gómez, J.A.; Baffa, O.; Dantas, N.O. Raman spectroscopy of very small $Cd_{1-x}Co_xS$ quantum dots grown by a novel protocol: direct observation of acoustic-optical phonon coupling. J. Raman Spectrosc. **2013**, *44*, 1022-1032, doi:10.1002/jrs.4324.

[43] Ardit, M.; Dondi, M.; Cruciani, G.; Zanelli, C. Tetrahedrally coordinated Co2+in oxides and silicates: Effect of local environment on optical properties. Am. Mineral. **2014**, *99*, 1736-1745, doi:10.2138/am.2014.4877.

[44] Tanabe, Y.; Sugano, S. On the Absorption Spectra of Complex Ions. J. Phys. Soc. Japan **1954**, *9*, 753-766, doi:10.1143/JPSJ.9.753.

Section 3

Application

Chapter 8

Doped Semiconductor Nanocrystals: Development and Applications

Anielle C.A. Silva, Eliete A. Alvin, Francisco R.A. dos Santos, Samanta L.M. de Matos, Jerusa M. de Oliveira, Alessandra S. Silva, Éder V. Guimarães, Mirella S. Vieira, Eurípedes A. da Silva Filho, Ricardo S. Silva, Lucas Anhezini, Nilvanira D. Tebaldi and Noelio O. Dantas

Abstract

This chapter aims to show significant progress that our group has been developing and the applications of several doped semiconductor nanocrystals (NCs), as nanopowders or embedded in glass systems. Depending on the type of dopant incorporated in the nanocrystals, the physical, chemical, and biological properties can be intensified. However, it can also generate undesired toxic effects that can potentially compromise its use. Here we present the potential of zinc oxide NCs doped with silver (Ag), gold (Au), and magnesium (Mg) ions to control bacterial diseases in agriculture. We have also performed biocompatibility analysis of the pure and Ag-doped sodium titanate ($Na_2Ti_3O_7$) NCs in *Drosophila*. The doped nanocrystals embedded in glassy systems are chrome (Cr) or copper (Cu) in ZnTe and Bi_2Te_3 NCs for spintronic development nanodevices. Therefore, we will show several advantages that doped nanocrystals may present in the technological and biotechnological areas.

Keywords: doping, semiconductor nanocrystals, agriculture, biological effects, development, glass system, nanopowders

1. Introduction

The development of new nanocrystals made from the doping of ions in semiconductors creates interesting physical–chemical properties and biological effects. The nanopowders are aiming at agricultural applications, and doped nanocrystals embedded in the glass system can be used in spintronics applications.

Bacterial diseases of plants occur in every place that is reasonably moist or warm, and they affect all kinds of plants. Bacterial diseases are prevalent and severe in the humid tropics, but they may be extremely destructive anywhere under favorable environmental conditions [1]. Control bacterial diseases in agriculture are complex because the few registered chemical products and the nanoparticles or nanocrystals emerge as an innovative method in disease management. Zinc-oxide (ZnO) nanoparticles or nanocrystals are among the most commonly used [2–5].

IntechOpen

ZnO, ZnS, or MgO nanoparticles have previously been used to control plant diseases caused by *Liberibacter crescens* [6]. *Xanthomonas alfalfae subsp. citrumelonis* and *Pseudomonas syringae* [7], *Xanthomonas perforans* [8], *Xanthomonas campestris pv. Campestris* [9], *Pantoea ananatis* [10, 11], *Xanthomonas axonopodis pv. Citri* [12], *Xanthomonas oryzae pv. Oryzae* [13], and *Xanthomonas citri subsp. Citri* [14]. ZnO nanocrystals (NCs) may be doped with various elements, such as noble metals or transition metals, to increase their bactericidal effect. Doping is a process that consists of adding new elements to the nanoparticle's structure and changing their chemical characteristics [2]. Therefore, in this chapter book, we will show the results of ZnO NCs doped with silver (Ag), gold (Au), and magnesium (Mg) ions to control bacterial diseases in agriculture.

The fruit fly *Drosophila melanogaster* is a well-established model organism in various areas of science, including nanotoxicology [15]. The fruit fly also has 77% of the conserved genes related to human diseases [16] and considerable similarities with humans in different physiological mechanisms [17].

Several mutant lines for a broad range of human diseases are available in this model, besides its low cost and easy maintenance in the laboratory, in addition to a short life cycle, when compared to other model organisms such as fishes and mammals. Taken together, these characteristics make *Drosophila* a valuable model for studies that evaluate long-term and developmental effects in nanotoxicology [15]. Here we present results regarding the biocompatibility analysis of the pure and Ag-doped $Na_2Ti_3O_7$ nanocrystals in *Drosophila*.

The doped nanocrystals can be embedded in glassy systems, allowing for various applicability in devices. $Zn_{1-x}A_xTe$ (A = Cr; Cu) nanocrystals (NCs) have been one investigated diluted magnetic semiconductor (DMS) system, due mainly to their strengthening *sp-d* exchange interactions with increasing A-doping concentration [18]. Cr and Cu-doped NCs simultaneously exhibit semiconductor and magnetic properties that may allow more diverse technological applications than undoped semiconductors [19, 20]. In this context, we present a very effective method for the growth of Cr^{2+} and Cu^{2+} ions-doped ZnTe NCs in a glass system ($65P_2O_5 \cdot 14ZnO \cdot 1Al_2O_3 \cdot 10BaO \cdot 10PbO$ (mol %), named PZABP) using the fusion nucleation method, as descriptions in Refs. [21, 22].

Bi_2Te_3 semiconductors at the nanoscale are highly performing materials for thermoelectric and promising applications as topological insulators [23]. These nano-semiconductors' physical and chemical properties can enhance and perform new features based on quantum behavior and the electronic structure's doping [24–26]. The synthesis of Bi_2Te_3 NCs in diamagnetic host glasses allows the samples' high chemical stability. During fusion, Cr ions can incorporate into these systems allowing possible applications in the manufacture of magneto-optical devices [25–27]. Therefore, the long-range magnetic properties generated by the domain of the Cr ion doping spins, in addition to the insulating topological states of the Bi_2Te_3 semiconductor NCs, have aroused great interest in the scientific community for the development of spintronic nanodevices [24, 27, 28]. Thus, we will show some results of Cr doped Bi_2Te_3 NCs.

Therefore, in this chapter, we show doped nanocrystals' results in powdered or embedded glass systems aiming at several applications.

2. Nanocrystals in powder or embedded in glass systems

Depending on how these nanocrystals are, for example, powdered or embedded in glass systems, the applicability is diverse. Thus, in applications in agronomy, dental, or biology, these doped nanocrystals must be in powder to be dispersed or not in solutions. In applications such as spintronics, the doped nanocrystals must be embedded in a thermally and chemically stable system, such as a glass system. We will comment on these peculiarities and advantages of each doped nanocrystal in the following sections.

2.1 Synthesis of Nanopowders and nanocrystals embedded in glassy matrices

2.1.1 Synthesis of Nanopowders

The pure and doped ZnO NCs were synthesized by coprecipitation by reference [29]. Pure and doped sodium titanate ($Na_2Ti_3O_7$) were synthesized by reference [30].

2.1.2 Synthesis of Cr or Cu-doped ZnTe nanocrystals embedded in glass matrix

The PZABP glass matrix with a nominal composition of $65P_2O_5 \cdot 14ZnO \cdot 1Al_2O_3 \cdot 10BaO \cdot 10PbO$ (mol %) adding 2Te (wt %), and Cr or Cu at doping x content varying with Zn content from 0 to 10 (wt %), were synthesized by fusion in alumina crucibles at 1300 °C for 30 minutes. These melted mixtures were quickly cooled to room temperature forming a glass system doped with the precursor ions needed for nanoparticle growth. Next, the glass samples were thermally annealed at 500 °C for 10 hours to enhance the diffusion of Zn^{2+}, Cr^{2+} or Cu^{2+} and Te^{2-} ions throughout the host PZABP matrix and induce the growth of $Zn_{1-x}Cr_xTe/Zn_{1-x}Cu_xTe$ NCs. The physical properties of the glass samples were studied by optical absorption (OA), recorded with a model UV-3600 Shimadzu UV–VIS–NIR spectrometer, operating between 190 and 3300 nm; XRD patterns were recorded using a XRD-6000 Shimadzu diffractometer equipped with monochromatic $CuK_{\alpha 1}$ radiation (k = 1.54056 Å) and set to a resolution of 0.02; Transmission electron micrographs (TEM JOEL, JEM-2100, 200 kV) and EPR measurements at temperature of 10 K were performed with a high sensitivity Bruker-EMX spectrometer operating at X-band (9.4 GHz) microwave frequency.

2.1.3 Synthesis of Cr-doped Bi_2Te_3 nanocrystals embedded in glass matrix

$Bi_{2-x}Cr_xTe_3$ NCs were synthesized by the fusion method in a borosilicate glass matrix with the following nominal composition: SNAB–$45SiO_2 \cdot 30Na_2CO_3 \cdot 5Al_2O_3 \cdot 20B_2O_3$ (mol %), 2% (of the weight of the glass matrix) of Te and Bi_2O_3, with nominal Cr content of x (x = 0.00, 0.01, and 0.05) as a function of bismuth concentration. The powdered glass and NC precursors were mixed together and melted in an alumina crucible at 1200 °C for 30 min and then rapidly cooled to room temperature, permitting the formation of Cr-doped Bi_2Te_3 NCs.

2.1.4 Synthesis of Cr-doped Bi_2Te_3 nanocrystals embedded in glass matrix

$Bi_{2-x}Cr_xTe_3$ NCs were synthesized by the fusion method in a borosilicate glass matrix with the following nominal composition: SNAB–$45SiO_2 \cdot 30Na_2CO_3 \cdot 5Al_2O_3 \cdot 20B_2O_3$ (mol %), 2% (of the weight of the glass matrix) of Te and Bi_2O_3, with nominal Cr content of x (x = 0.00, 0.01, and 0.05) as a function of bismuth concentration. The powdered glass and NC precursors were mixed together and melted in an alumina crucible at 1200 °C for 30 min and then rapidly cooled to room temperature, permitting the formation of Cr-doped Bi_2Te_3 NCs.

2.1.5 Nanocrystals for the control plant bacterial disease

To evaluate the growth inhibition zone of *Xanthomonas campestris pv. campestris in vitro,* a basic layer of 2% agar-water medium and semi-solid nutrient medium (0.8%)

with 10% of the bacterial suspension (10^8 CFU/mL) was added to Petri dishes. Seven disks of sterile filter paper (6 mm) were embedded with 10 μL of solution Ag-doped ZnO NCs, and Au-doped ZnO NCs, at 100; 10; 1; 0.1; and 0.01 mg/mL, and incubated at 28 °C for 48 hours. The diameter of the inhibition zone was measured.

In the greenhouse, tomato plants (three- to four-leaf stage) were sprayed with Mg-doped ZnO NCs at 2.5 mg/mL, and three days later, the plants were inoculated with a *Xanthomonas gardneri*-bacterial suspension (10^9 CFU/mL). The severity of the disease was analyzed.

2.1.6 In vivo biocompatibility analysis in Drosophila melanogaster

We have performed a developmental assay to evaluate whether Na_2TiO_7 and Na_2TiO_7:Ag could present any toxicity *in vivo*. Adult males and females were conditioned and kept in oviposition plates for six hours. After 24 hours of controlled oviposition, first instar larvae (L1) were carefully transferred (six replicates per concentration containing 30 larvae each) to standard *Drosophila* medium containing Na_2TiO_7 and Na_2TiO_7:Ag at the final concentration of 0.02 mg/mL (66.30 μmol/L) while control larvae were transferred to a standard culture medium. The animals developed through all larval stages during the following 4 days, while they actively fed until reaching the pupal stage. Animals that reached pupal stage were counted daily and it was possible to determine total pupation and daily pupation rate. We obtained the larval lethality rate by subtracting the number of pupae from the initial larvae number. After pupal metamorphosis, the animals emerge as adults and were transferred to a standard media and monitored throughout its adult lives to perform the adult lifespan assay. Deaths were counted daily until all animals were dead.

2.2 Nanocrystals in powder: Development and applicability

In this section, we will show the results of ZnO NCs doped with silver (Ag), gold (Au), and magnesium (Mg) ions to control bacterial diseases in agriculture. We will also present *in vivo* biocompatibility assays of pure and Ag-doped $Na_2Ti_3O_7$ nanocrystals in *Drosophila* development.

2.2.1 Nanocrystals application in agriculture to control bacterial diseases

The nanocrystals of ZnO in this work have a wurtzite structure, and silver (Ag), gold (Au), and magnesium (Mg) ions were doped in ZnO NCs, as shown in **Figure 1a**. The Ag or Au doped ZnO NCs inhibited *Xanthomonas campestris pv. camprestris* growth, at 100 and 10 mg/mL (**Figure 1b,c**). Other elements can also inhibit, such as, Mg ions that inhibited the growth of *Xanthomonas gardneri,* and reduced the severity of tomato bacterial spot (**Figure 1e**), and controlled the bacteria present in tomato seeds [31]. Therefore, nanotechnology could sustainably mitigate many challenges in disease management by reducing agrochemical use [32].

2.2.2 Biocompatibility In Vivo of nanocrystals

The crystalline structure of pure and silver (Ag) doped sodium titanate ($Na_2Ti_3O_7$) is shown in **Figure 2a**. The Ag ions replace sodium (Na) or titanium (Ti) ions in sodium titanate's crystalline structure.

In order to investigate whether $Na_2Ti_3O_7$ and Ag-doped $Na_2Ti_3O_7$ nanocrystals could be biocompatible, we performed a bioassay to evaluate the effects of these NCs during *Drosophila* development. Surprisingly, the animals that developed

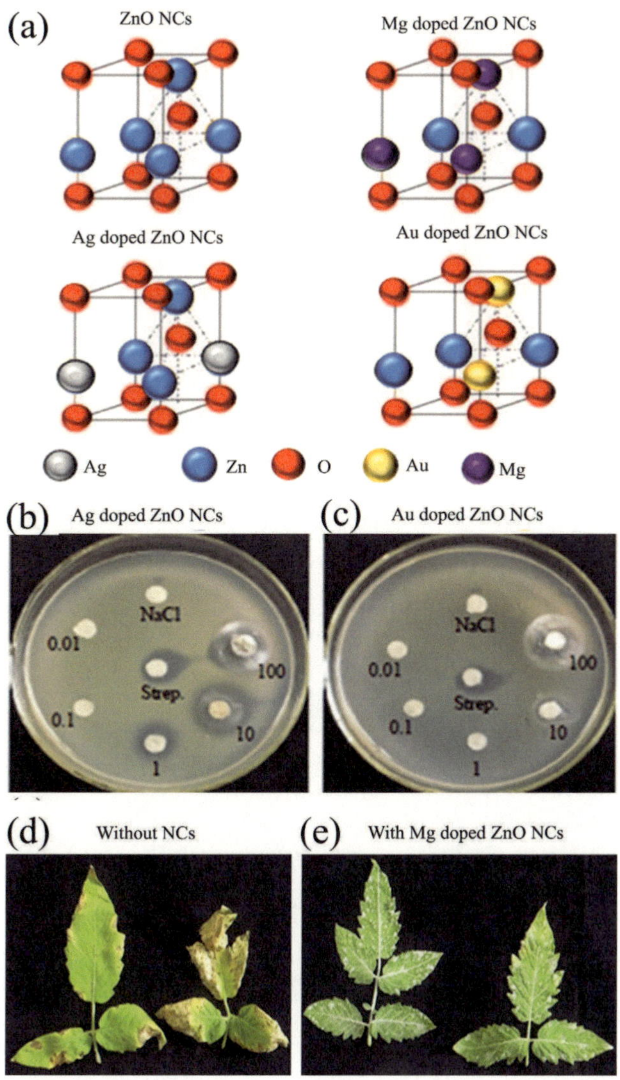

Figure 1.
(a) Wurtzite structure of ZnO, Ag-doped ZnO, Au doped ZnO, and Mg-doped ZnO NCs. Growth inhibition zone for Xanthomonas campestris pv. campestris *treated with (b) Ag-doped ZnO NCs and (c) Au doped ZnO NCs, at 100; 10; 1; 0.1; and 0.01 mg/mL. Strep. = streptomycin. Symptoms of bacterial spot-on tomato leaves, (d) caused by* Xanthomonas gardneri, *(e) with Mg-doped ZnO NCs showing disease control.*

at Ag-doped $Na_2Ti_3O_7$ NCs exhibit a lethality rate 12.8% lower when compared to animals that developed on Na_2TiO_7 NCs, suggesting that Ag doping was able to increase the $Na_2Ti_3O_7$ NCs biocompatibility *in vivo* by decreasing its toxicity (**Figure 2b**).

As observed in **Figure 3a**, there was no delay in the time the larvae took to reach the pupal stage when exposed to $Na_2Ti_3O_7$ and Ag-doped $Na_2Ti_3O_7$ when compared to control. We have also performed an adult lifespan assay to evaluate the effects of NCs exposure during larval development and pupal metamorphosis over the adult survival. Therefore, after pupal metamorphosis, the animals that emerged as adults were immediately separated and kept in vials with standard control medium. These animals were transferred to a new vial with fresh standard medium every five days. The number of deaths for each experimental group was recorded daily until all individuals were dead. The lifetime of individuals that have developed in media

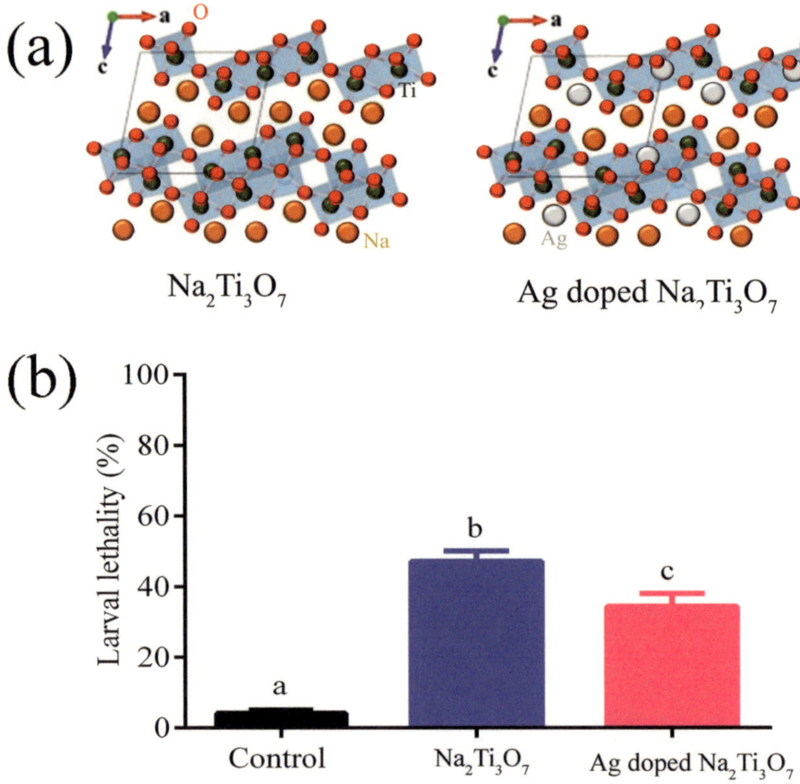

Figure 2.
(a) Crystalline structure of pure and silver (Ag) doped sodium titanate ($Na_2Ti_3O_7$). (b) Larval lethality following NCs exposure. As observed $Na_2Ti_3O_7$ exposure caused a significant lethality rate during larval development, which was partially rescued by the Ag-doped $Na_2Ti_3O_7$.

containing $Na_2Ti_3O_7$ and Ag-doped $Na_2Ti_3O_7$ was compared to control animals that developed in a standard culture medium.

As shown in **Figure 3b** animals that developed in standard *Drosophila* medium containing $Na_2Ti_3O_7$ showed 20 days decrease in lifespan when compared to control. Surprisingly, the emerged adult flies that developed in medium containing Ag-doped $Na_2Ti_3O_7$ showed a longer longevity even when compared to animals that developed in standard culture medium or medium containing $Na_2Ti_3O_7$ NCs. Therefore, our data suggests that Ag-doped $Na_2Ti_3O_7$ was able to increase the $Na_2Ti_3O_7$ NCs biocompatibility *in vivo* by decreasing its toxicity.

The toxicity of NCs, such as zinc oxide, titanium dioxide, magnetite, hydroxyapatite, and sodium titanate, is induced through the generation of reactive species and consequent oxidative stress [33–35]. The redox imbalance caused by NCs is capable of generating mitochondrial dysfunctions, inducing inflammatory responses, causing cytotoxicity and genotoxicity, in addition to altering the functioning of the sodium and potassium channels and consequent cell death [33]. Oxidative stress not only causes cell damage and protein oxidation but is also possibly responsible for altering the biosynthesis of hormones, such as ecdysone in insects. Ecdysone is a crucial hormone in the control of metamorphosis and ecdysis events in insects [36, 37]. Our data showed a high larval lethality, especially in animals exposed to $Na_2Ti_3O_7$, possibly generated by oxidative stress, which can also impair ecdysone biosynthesis, causing developmental problems. One of the forgotten properties of NCs is their antioxidant capacity, such as Ag and cerium oxide

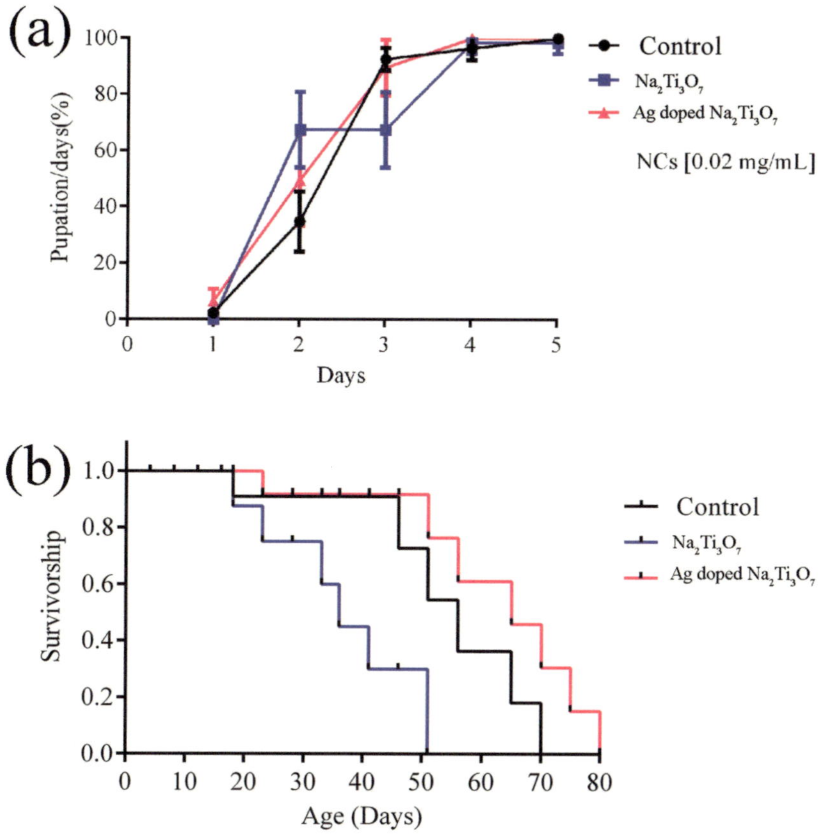

Figure 3.
(a) Daily pupation analysis of pure and Ag-doped $Na_2Ti_3O_7$ exposed animals. It is possible to observe that the exposure to pure and Ag-doped $Na_2Ti_3O_7$ Ag caused no delay in the transition from larva to pupa. (b) Lifespan analysis of pure and Ag-doped $Na_2Ti_3O_7$ exposed animals. It is shown that $Na_2Ti_3O_7$ exposure during development decreased adult longevity in about 20 days in comparison to control animals. However, the Ag-doped $Na_2Ti_3O_7$ not only rescued the animals survival but surprisingly increased it in 10 days in comparison to control animals.

nanoparticles (CONPs), and some NP of oxide that can even mimic an antioxidant molecule [38].

Therefore, we believe that the reduction in larval lethality observed for Ag-doped $Na_2Ti_3O_7$ NCs compared to $Na_2Ti_3O_7$ NCs can be explained by a lesser effect on the generation of reactive species, suggesting that the transition metal silver was sufficient to increase the biocompatibility of $Na_2Ti_3O_7$ NCs *in vivo*. However, additional studies are crucial to better understand the mechanisms by which development is influenced by nanocrystals.

2.3 Doped nanocrystals embedded in glass systems

This section will show the results of $Zn_{1-x}Cr_xTe$ NCs, $Zn_{1-x}Cu_xTe$ NCs, and $Bi_{2-x}Cr_xTe_3$ NCs embedded in glass systems aiming at spintronics applications.

2.3.1 Cr- and Cu-doped ZnTe nanocrystals embedded in glass matrix

Figure 4 presents TEM images (**a**) and X-ray diffraction (XRD) (**b**) of samples containing $Zn_{1-x}Cr_xTe$ NCs, with Cr-concentrations ranging from x = 0.00 to

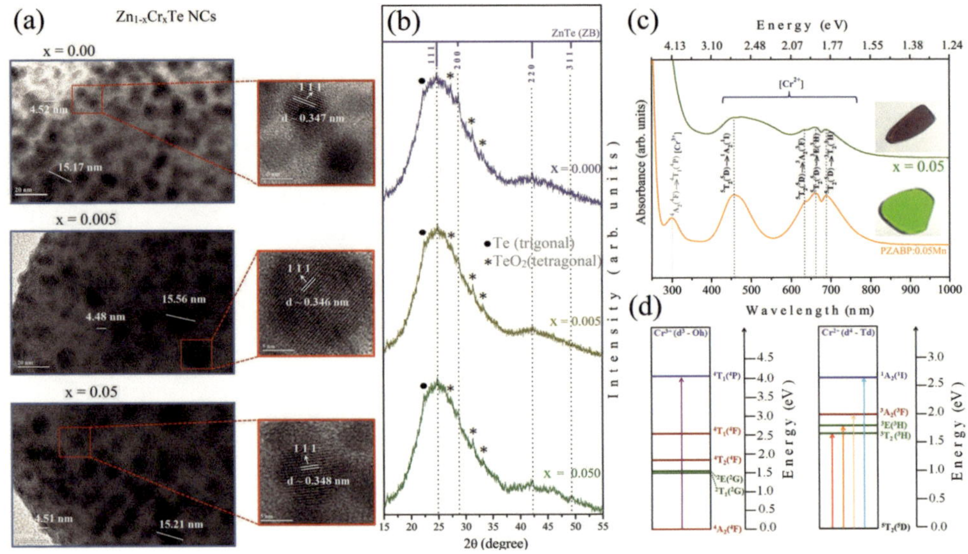

Figure 4.
TEM images (a) and XRD diffractograms (b) of samples containing $Zn_{1-x}Cr_xTe$ NCs, with Cr-concentrations ranging from x = 0.00 to x = 0.05; (c) OA spectra of glass samples containing $Zn_{1-x}Cr_xTe$ NCs, with concentration x = 0.05 and PZABP:0.05Cr; (d) energy diagram of Cr^{3+} and Cr^{2+} ions coordinated octahedral (oh) and tetrahedral (td).

$x = 0.05$. Already the **Figure 4(c)** shows OA spectra of glass samples containing $Zn_{1-x}Cr_xTe$ NCs, with concentration $x = 0.05$ and PZABP glassy matrix doped with 0.05 Cr (PZABP:0.05Cr). In **Figure 4(d)**, energy diagram (below) of Cr^{3+} and Cr^{2+} ions coordinated octahedral (Oh) and tetrahedral (Td) of the transitions observed [22].

The formation of $Zn_{1-x}Cr_xTe$ NCs was confirmed by Transmission Electron Microscopy (TEM) images. These images show two distinct groups of spherical NCs attributed to quantum dots (QDs) and bulk NCs, as we reported in our previous work [21]. The average diameters of these NCs are approximately D ~ 4.50 nm for QDs and D ~ 15.31 for bulk NCs. TEM images also show that the distance between their crystallographic planes of these does not vary with the Cr concentration. This suggests invariance in lattice parameter with the incorporation of Cr^{2+} ions. This result is expected, and is in accordance with what was observed from the XRD difractograms, because the ionic radii of Zn^{2+} (0.68 nm) and Cr^{2+} (0.73 nm) are very similar. Thus, XRD show that the position of the peaks, corresponding to (1 1 1), (2 0 0) (2 2 0) and (3 1 1), is the same for all samples. Optical absorption spectra confirm the substitutional incorporation of Cr^{2+} ions in the ZnTe semiconductor lattice, due the ${}^5T_2({}^5D) \rightarrow {}^1A_2({}^1I)$, ${}^5T_2({}^5D) \rightarrow {}^3A_2({}^3F)$, ${}^5T_2({}^5D) \rightarrow {}^3E({}^3H)$ and ${}^5T_2({}^5D) \rightarrow {}^3T_2({}^3H)$ spin forbidden absorption bands of Cr^{2+} ions [39], indicated in the energy diagram.

Figure 5 shows OA spectra and photographs (**a**) of the PZABP template and of $Zn_{1-x}Cu_xTe$ NC samples embedded in this template at different Cu-doping contents: x varying from 0.000 to 0.100. **Figure 5(b)** shows TEM images for sample with x = 0.05 (**b**) and EPR spectra for sample with x = 0.10 **Figure 5(c)** [21]. A redshift in the OA bands assigned to the QDs as Cu concentration increases, shifting from 3.10 eV (400 nm) to 2.95 eV (420 nm), when x ranges from 0.00 to 0.10. This decrease in the band gap energy with increase of transition metal (Cu) doped II–IV compound semiconductors can be best understand in terms of *sp–d* spin exchange interaction between band electrons and the localized *d* electrons of the transition metal ions substituting the cations [40]. TEM images confirm the formation of

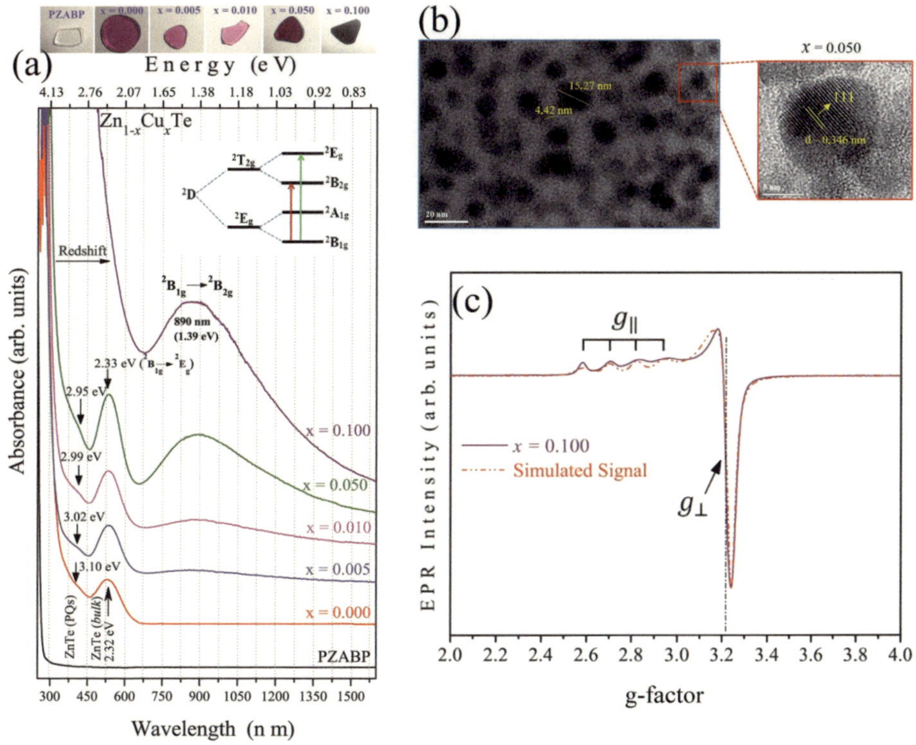

Figure 5.
OA spectra and photographs (a) of the PZABP template and of $Zn_{1-x}Cu_xTe$ NC samples embedded in this template at different Cu-doping contents: x varying from 0.00 to 0.100. TEM images for sample with x = 0.05 (b) and EPR spectra for sample with x = 0.10 (c).

$Zn_{1-x}Cu_xTe$ NCs, with interplanar distance around d ~ 0.346 nm. Electron paramagnetic resonance spectra (EPR) confirm the substitutional incorporation, due to hyperfine transitions characteristic Cu^{2+} ions with d^9 configuration.

2.3.2 Cr-doped Bi_2Te_3 nanocrystals embedded in glass systems

The UV–VIS optical absorption (OA) spectra shown in **Figure 6(a)** for the host glass matrix SNAB- $45SiO_2{\cdot}30Na_2CO_3{\cdot}5Al_2O_3{\cdot}20B_2O_3$ (mol%) of the $Bi_{2-x}Cr_xTe_3$ NCs with xCr molar fraction (x = 0.00; 0.01; 0.05) provides strong evidence of the formation and incorporation of Cr^{3+} ions in orthorhombic sites of Bi_2Te_3 NCs. The SNAB matrix with a band gap of approximately 4 eV [26], is an ideal template for observing d-d and excitonic transitions in $Bi_{2-x}Cr_xTe_3$ NCs as shown by the spectrum's bottom line (black) with no band in the visible region.

Bi_2Te_3 is a V - VI semiconductor that presents a narrow band gap of 0.13 eV in bulk form at room temperature [41]. The tail of the band attributed to the absorption of the charge carriers (electron–hole pair) shows the result of the nucleation and formation of the Bi_2Te_3 NCs with a confinement energy around 3.10 eV. The slight blueshift observed for the $Bi_{2-x}Cr_xTe_3$ NCs (x = 0.01; 0.05) is due to the strong sp-d exchange interactions between Bi_2Te_3 (sp) excitons and the electrons of the unoccupied $3d^3$ orbitals of the Cr^{3+} ion. This orbital coupling modifies the optical properties of the intrinsic semiconductor, proportional to the increase in Cr content. Finally, the sharpness of the bands of increasing intensity observed in the visible spectral region is due to the 3d-3d electronic transitions of the Cr ions [26, 42].

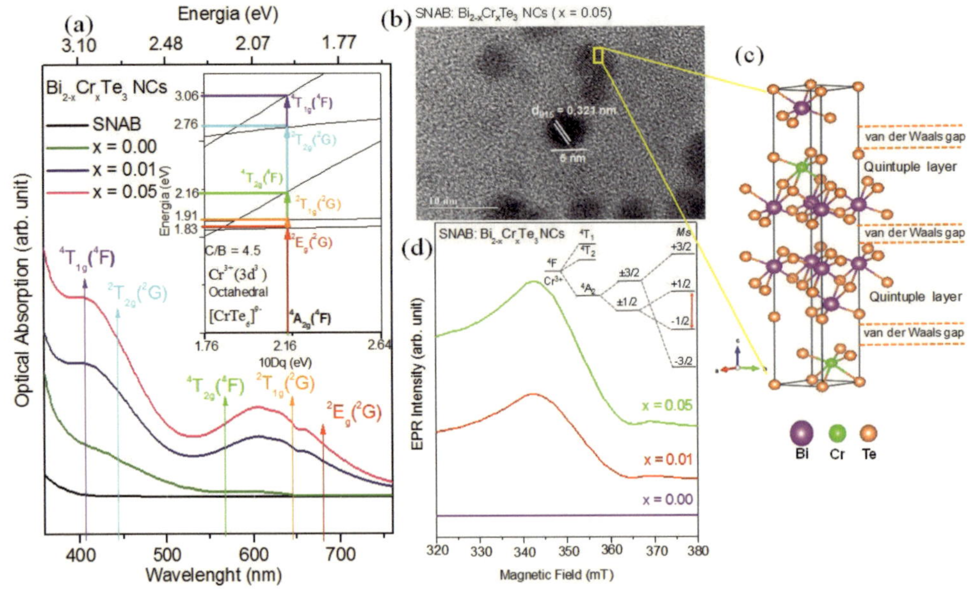

Figure 6.
(a) Optical absorption spectra at room temperature of $Bi_{2-x}Cr_xTe_3$ NCs (x = 0.00; 0.01; 0.05) embedded in SNAB glass matrix. For comparison purposes, the absorption spectrum of the SNAB glass matrix represents on the black bottom line. The inset shows the Tanabe-Sugano diagram d^3 of octahedral symmetry (C/B = 4.5) for the $[CrTe_6]^{9-}$ complex and the respective spin allowed and forbidden transitions indicated on the energy 10 Dq = 2.16 eV. (b) TEM image of $Bi_{2-x}Cr_xTe_3$ NCs (x = 0.05) embedded in SNAB glass. (c) Details of the quintuple layer and the van der Waals gap in the Bi_2T_3 hexagonal unit cell with the substitutional doping of Bi^{3+} ions by Cr^{3+} in distorted octahedral sites. (d) EPR spectra in the X band, at 300 K for NCs of $Bi_{2-x}Cr_xTe_3$ NCs (x = 0.00; 0.01; 0.05) embedded in the SNAB glass matrix. The inset shows the split diagram of the energy states of the system.

The energy states identified in the OA spectrum of figure x (a) belong to the spin allowed and forbidden d-d transitions: $^4A_2(^4F) \rightarrow {}^2E_g\,(^2G)$ (1.83 eV), $^4A_2(^4F) \rightarrow {}^2T_1(^2G)$ (1.91 eV), $^4A_{2g}\,(^4F) \rightarrow {}^4T_{2g}\,(^4F)$ (2.16 eV), $^4A_{2g}\,(^4F) \rightarrow {}^2T_{2g}\,(^2G)$ (2.76 eV) and $^4A_2\,(^4F) \rightarrow {}^4T_1\,(^4F)$ (3.06 eV). These transitions are in accordance with a Tanabe-Sugano diagram d^3 of octahedral symmetry for C/B = 4.5 (see inset in **Figure 6a**) [43]. The results are typical of inter-electronic repulsion parameters (Racah) B = 0.088 eV in a crystal field strength 10 Dq (Δ) = 2.16 eV of Cr^{3+} ions in coordinated octahedral sites of Te ($[CrTe_6]^{9-}$) ligands [42, 44].

The exciton Bohr radius of approximately 50 nm for Bi_2Te_3 bulk [45] makes the semiconductor subject to strong quantum confinements. **Figure 6(b)** shows the TEM image of the SNAB glass matrix host of the $Bi_{2-x}Cr_xTe_3$ NCs (x = 0.05). The nanocrystal's 5 nm size confirms the formation of $Bi_{2-x}Cr_xTe_3$ quantum dots due to the strong quantum confinement of the semiconductor structure.

The quantum size of the $Bi_{2-x}Cr_xTe_3$ NCs does not change with the increase of Cr incorporation in the samples. In this way, the structure preserves due to the non-saturation of the molar fraction of Cr doping in the $Bi_{2-x}Cr_xTe_3$ NCs. The interplanar distance d_{015} = 0.321 nm is evidence of Tellurobismuthite's hexagonal crystalline structure [27, 41]. **Figure 6(c)** shows the hexagonal unit cell $R\underline{3}m - D_{3d}^5$ [41] of the Cr-doped Bi_2Te_3 NCs with the atomic arrangement of monoatomic planes Te - Cr - Te - Bi - Te and Te - Bi - Te - Bi - Te in terms of quintuple layers linked by weak van der Waals interactions [41]. The substitutional doping of Cr^{3+} ions with a smaller ionic radius (0.53 Å) in relation to Bi^{3+} (1.03 Å) distorts the environment of octahedral symmetry [42].

Figure 6(d) shows the EPR measurements in the X band, at 300 K for the $Bi_{2-x}Cr_xTe_3$ NCs (x = 0.00; 0.01; 0.05) samples embedded in the SNAB glass matrix. The EPR spectrum for Bi_2Te_3 not doped does not show any signal due to the absence of doping ions. However, the central signal characteristic of a Ms. = ± ½ transition typical of a fine structure line results from the interaction between the electronic (S = 3/2) and nuclear (I = 0) spins of Cr^{3+} ions in an octahedral crystal field. The inset in **Figure 6(d)** shows the split diagram of the energy states involved in the system. The increasing molar fraction of Cr ions in the $Bi_{2-x}Cr_xTe_3$ results in greater dipole–dipole interaction and, consequently, an increase in the observed RPE signal's intensity [26].

Therefore, the long-range magnetic properties generated by the domain of the Cr ion doping spins, in addition to the insulating topological states of the Bi_2Te_3 semiconductor NCs, have aroused great interest in the scientific community for the development of spintronic nanodevices [24, 25, 27, 28].

3. Conclusion

Therefore, this chapter showed the development and applications of several doped semiconductor nanocrystals, as nanopowders or embedded in glass systems. Doped Nanocrystals show good potential to control plant diseases as controlling bacterial diseases on field crops is complex. We also demonstrate that depending on the ion incorporated in the nanocrystal structure, the biocompatibility could be improved. Additionally, we show magnetic properties generated by the domain of the Cu or Cr ions doping spins, in addition to semiconductor nanocrystals embedded in glass systems, for the development of spintronic nanodevices.

Acknowledgements

This work was supported by grants of CNPq, CAPES, and FAPEMIG.

Conflict of interest

The authors declare no conflict of interest.

Author details

Anielle C.A. Silva[1,2*], Eliete A. Alvin[1,2], Francisco R.A. dos Santos[1,3], Samanta L.M. de Matos[1,3], Jerusa M. de Oliveira[3], Alessandra S. Silva[4], Éder V. Guimarães[4], Mirella S. Vieira[1,5], Eurípedes A. da Silva Filho[6], Ricardo S. Silva[4], Lucas Anhezini[3], Nilvanira D. Tebaldi[7] and Noelio O. Dantas[1*]

1 Laboratory of New Nanostructured and Functional Materials, Physics Institute, Federal University of Alagoas, Maceió, Alagoas, Brazil

2 Programa de Pós-Graduação da Rede Nordeste de Biotecnologia (RENORBIO), Federal University of Alagoas, Maceió, Alagoas, Brazil

3 Laboratory for in vivo Toxicity, Institute of Biological Sciences and Health, Federal University of Alagoas, Maceió, Alagoas, Brazil

4 Instituto de Ciências Exatas, Naturais e Educação (ICENE), Departamento de Física, Federal University of Triângulo Mineiro, Minas Gerais, Brazil

5 Faculty of Electrical Engineering, Federal University of Uberlândia, Uberlândia, Brazil

6 Laboratory of Genetic and Applied Microbiology, Institute of Biological Sciences and Health, Federal University of Alagoas, Maceió, Alagoas, Brazil

7 Laboratory of Plant Bacteriology, Institute of Agricultural Sciences, Federal University of Uberlândia, Uberlândia, Brazil

*Address all correspondence to: acalmeida@fis.ufal.br and noelio@fis.ufal.br

IntechOpen

© 2021 The Author(s). Licensee IntechOpen. This chapter is distributed under the terms of the Creative Commons Attribution License (http://creativecommons.org/licenses/by/3.0), which permits unrestricted use, distribution, and reproduction in any medium, provided the original work is properly cited. CC BY

References

[1] Agrios, G.N. *Plant pathology*; 5th ed.; 2005;

[2] Silva, A.; Zóia, M.A.P.; Correia, L.I.V.; Azevedo, F.V.P.V.; Paula, A.T. de; Maia, L.P.; Carvalho, L.S. de; Carvalho, L.N.; Costa, M.P.C.; Giaretta, L.C.; et al. Biocompatibility of Doped Semiconductors Nanocrystals and Nanocomposites. In *Cytotoxicity*; InTech, 2018.

[3] Batista, E.A.; Silva, A.C.A.; de Lima, T.K.; Guimarães, E.V.; da Silva, R.S.; Dantas, N.O. Effect of the location of Mn2+ ions in the optical and magnetic properties of ZnO nanocrystals. J. Alloys Compd. **2021**, *850*, 156611, doi:10.1016/j.jallcom.2020.156611.

[4] Souza, G.L. de; Moura, C.C.G.; Silva, A.C.A.; Marinho, J.Z.; Silva, T.R.; Dantas, N.O.; Bonvicini, J.F.S.; Turrioni, A.P. Effects of zinc oxide and calcium–doped zinc oxide nanocrystals on cytotoxicity and reactive oxygen species production in different cell culture models. Restor. Dent. Endod. **2020**, *45*, 54, doi:10.5395/rde.2020.45.e54.

[5] Dantas, N.O.; Damigo, L.; Qu, F.; Cunha, J.F.R.; Silva, R.S.; Miranda, K.L.; Vilela, E.C.; Sartoratto, P.P.C.; Morais, P.C. Raman investigation of ZnO and Zn1−xMnxO nanocrystals synthesized by precipitation method. J. Non. Cryst. Solids **2008**, *354*, 4827-4829, doi:10.1016/j.jnoncrysol.2008.04.051.

[6] Naranjo, E.; Merfa, M. V.; Santra, S.; Ozcan, A.; Johnson, E.; Cobine, P.A.; De La Fuente, L. Zinkicide is a ZnO-based nanoformulation with bactericidal activity against liberibacter crescens in batch cultures and in microfluidic chambers simulating plant vascular systems. Appl. Environ. Microbiol. **2020**, *86*, 1-18, doi:10.1128/AEM.00788-20.

[7] Maxwell, T.J.; Rajasekaran, P.; Young, M.; Schaff, M.; Heetai, R.; Santra, S. Non-phytotoxic zinc based nanoparticle adjuvant for improving rainfastness and sustained release of streptomycin. *Environ. Nanotechnology, Monit. Manag.* **2020**, *14*, 100355, doi:10.1016/j.enmm.2020.100355.

[8] Liao, Y.-Y.; Strayer-Scherer, A.L.; White, J.; Mukherjee, A.; De La Torre-Roche, R.; Ritchie, L.; Colee, J.; Vallad, G.E.; Freeman, J.H.; Jones, J.B.; et al. Nano-Magnesium Oxide: A Novel Bactericide Against Copper-Tolerant *Xanthomonas perforans* Causing Tomato Bacterial Spot. *Phytopathology®* **2019**, *109*, 52-62, doi:10.1094/PHYTO-05-18-0152-R.

[9] Zancan, N. Terapia fotodinâmica e nanopartículas no controle de Xanthomonas campestris pv. campestris in vitro e em sementes de canola naturalmente contaminadas, Universidade Federal de Uberlândia, 2018.

[10] Mota, R.P. da Agressividade, caracterização molecular de isolados de Pantoea ananatis e nanoparticulas no controle da bactéria na cultura do milho, Universidade Federal de Uberlândia: Uberlândia, 2019.

[11] Mamede, M.C. Detecção de *Pantoea ananatis* em sementes de milho e nanopartículas no controle da bactéria *in vitro*, Universidade Federal de Uberlândia: Uberlândia, 2018.

[12] Ballottin, D.; Fulaz, S.; Cabrini, F.; Tsukamoto, J.; Durán, N.; Alves, O.L.; Tasic, L. Antimicrobial textiles: Biogenic silver nanoparticles against Candida and Xanthomonas. Mater. Sci. Eng. C **2017**, *75*, 582-589, doi:10.1016/j.msec.2017.02.110.

[13] Liang, Y.; Yang, D.; Cui, J. A graphene oxide/silver nanoparticle composite as a novel agricultural antibacterial agent against

Xanthomonas oryzae pv. oryzae for crop disease management. *New J.* Chem. **2017**, *41*, 13692-13699, doi:10.1039/c7nj02942j.

[14] Graham, J.H.; Johnson, E.G.; Myers, M.E.; Young, M.; Rajasekaran, P.; Das, S.; Santra, S. Potential of Nano-Formulated Zinc Oxide for Control of Citrus Canker on Grapefruit Trees. Plant Dis. **2016**, *100*, 2442-2447, doi:10.1094/PDIS-05-16-0598-RE.

[15] Ong, C.; Yung, L.Y.L.; Cai, Y.; Bay, B.H.; Baeg, G.H. Drosophila melanogaster as a model organism to study nanotoxicity. Nanotoxicology **2015**, *9*, 396-403, doi:10.3109/17435390.2014.940405.

[16] Reiter, L.T.; Potocki, L.; Chien, S.; Gribskov, M.; Bier, E. A systematic analysis of human disease-associated gene sequences in Drosophila melanogaster. Genome Res. **2001**, *11*, 1114-1125, doi:10.1101/gr.169101.

[17] Pandey, U.B.; Nichols, C.D. Human disease models in Drosophila melanogaster and the role of the fly in therapeutic drug discovery. Pharmacol. Rev. **2011**, *63*, 411-436, doi:10.1124/pr.110.003293.

[18] Counio, G.; Gacoin, T.; Boilot, J.P. Synthesis and photoluminescence of Cd1-xMnxS ($x \leq 5\%$) nanocrystals. J. Phys. Chem. B **1998**, *102*, 5257-5260, doi:10.1021/jp980511w.

[19] Hu, T.; Zhang, M.; Wang, S.; Shi, Q.; Cui, G.; Sun, S. CdS:Co diluted magnetic semiconductor nanocrystals: Synthesis and ferromagnetism study. CrystEngComm **2011**, *13*, 5646-5649, doi:10.1039/c1ce05593c.

[20] Ogale, S.B. Dilute Doping, Defects, and Ferromagnetism in Metal Oxide Systems. Adv. Mater. **2010**, *22*, 3125-3155, doi:10.1002/adma.200903891.

[21] Silva, A.S.; Pelegrini, F.; Figueiredo, L.C.; de Souza, P.E.N.; Morais, P.C.; Dantas, N.O. Effects of Cu2+ ion incorporation into ZnTe nanocrystals dispersed within a glass matrix. J. Alloys Compd. **2018**, *749*, 681-686, doi:10.1016/j.jallcom.2018.03.290.

[22] Silva, A.S.; Lourenço, S.A.; Da Silva, M.A.T.; Dantas, N.O. Optical properties of Cr-doped Zn1-xMnxTe semimagnetic nanocrystals. Appl. Phys. Lett. **2018**, *112*, doi:10.1063/1.5016382.

[23] Qi, X.; Ma, W.; Zhang, X.; Zhang, C. Raman characterization and transport properties of morphology-dependent two-dimensional Bi 2 Te 3 nanofilms. Appl. Surf. Sci. **2018**, *457*, 41-48, doi:10.1016/j.apsusc.2018.06.142.

[24] Irfan, S.; Luo, J.; Ping, F.; Zhuanghao, Z. Theoretical and experimental investigation of magnetic properties of iodine and cerium Co-doped Bi2Te3 nanoparticles. J. Mater. Res. Technol. **2020**, *9*, 13893-13901, doi:10.1016/j.jmrt.2020.09.086.

[25] Silva, R.S.; Gualdi, A.J.; Zabotto, F.L.; Cano, N.F.; Silva, A.C.A.; Dantas, N.O. Weak ferromagnetism in Mn2+ doped Bi2Te3 nanocrystals grown in glass matrix. J. Alloys Compd. **2017**, *708*, 619-622, doi:10.1016/j.jallcom.2017.03.066.

[26] Araujo, C.M.; Mikhail, H.D.; Guimarães, E. V.; Rastrello, L.R.; Cano, N.F.; Silva, A.C.A.; Dantas, N.O.; Silva, R.S. Optical, structural and magnetic characterization of Bi2−xCrxTe3 nanocrystals in oxide glass. *Mater.* Chem. Phys. **2020**, *241*, doi:10.1016/j.matchemphys.2019.122323.

[27] Silva, R.S.; Mikhail, H.D.; Pavani, R.; Cano, N.F.; Silva, A.C.A.; Dantas, N.O. Synthesis of diluted magnetic semiconductor Bi2-xMnxTe3 nanocrystals in a host glass matrix. J. Alloys Compd. **2015**, *648*, 778-782, doi:10.1016/j.jallcom.2015.07.045.

[28] Fert, A.; Van Dau, F.N. Spintronics, from giant magnetoresistance to

magnetic skyrmions and topological insulators. Comptes Rendus Phys. 2019, *20*, 817-831.

[29] Silva, A.; Zóia, M.A.P.; Correia, L.I.V.; Azevedo, F.V.P.V.; Paula, A.T. de; Maia, L.P.; Carvalho, L.S. de; Carvalho, L.N.; Costa, M.P.C.; Giaretta, L.C.; et al. Biocompatibility of Doped Semiconductors Nanocrystals and Nanocomposites. In *Cytotoxicity*; InTech, 2018.

[30] Umek, P.; Gloter, A.; Navio, C.; Bittencourt, C. Synthesis and characterization of sodium titanate and TiO2 nanostructures loaded with silver nanoparticles. In Proceedings of the AIP Conference Proceedings; American Institute of PhysicsAIP, 2011; Vol. 1415, pp. 24-27.

[31] Fraga, F. Controle de Xanthomonas spp. com nanocristais de ZnO puros e dopados no tomateiro, Universidade Federal de Uberlândia, 2020.

[32] Elmer, W.; White, J.C. The future of nanotechnology in plant pathology. Annu. Rev. Phytopathol. 2018, *56*, 111-133.

[33] Manke, A.; Wang, L.; Rojanasakul, Y. Mechanisms of nanoparticle-induced oxidative stress and toxicity. Biomed Res. Int. **2013**, *2013*, doi:10.1155/2013/942916.

[34] Ajdary, M.; Moosavi, M.A.; Rahmati, M.; Falahati, M.; Mahboubi, M.; Mandegary, A.; Jangjoo, S.; Mohammadinejad, R.; Varma, R.S. Health concerns of various nanoparticles: A review of their in vitro and in vivo toxicity. Nanomaterials **2018**, *8*, 1-28, doi:10.3390/nano8090634.

[35] Singh, A.; Raj, A.; Shah, P.; Agrawal, N. Synthesis and Characterization of Nanoparticles Used in Consumer Products. In *Toxicology of Nanoparticles: Insights from Drosophila*; Springer Singapore, 2020; pp. 1-27.

[36] Uryu, O.; Ameku, T.; Niwa, R. Recent progress in understanding the role of ecdysteroids in adult insects: Germline development and circadian clock in the fruit fly Drosophila melanogaster. Zool. Lett. **2015**, *1*, 1-9, doi:10.1186/s40851-015-0031-2.

[37] Yamanaka, N.; Marqués, G.; O'Connor, M.B. Vesicle-Mediated Steroid Hormone Secretion in Drosophila melanogaster. Cell **2015**, *163*, 907-919, doi:10.1016/j.cell.2015.10.022.

[38] Kumar, H.; Bhardwaj, K.; Nepovimova, E.; Kuča, K.; Dhanjal, D.S.; Bhardwaj, S.; Bhatia, S.K.; Verma, R.; Kumar, D. Antioxidant functionalized nanoparticles: A combat against oxidative stress. Nanomaterials **2020**, *10*, 1-31, doi:10.3390/nano10071334.

[39] Nitsuk, Y.A. Energy states of a Cr2+ ion in ZnSe crystals. Semiconductors **2013**, *47*, 736-739, doi:10.1134/S1063782613060183.

[40] Rusu, G.I.; Diciu, M.; Pîrghie, C.; Popa, E.M. Structural characterization and optical properties of ZnSe thin films. Appl. Surf. Sci. **2007**, *253*, 9500-9505, doi:10.1016/j.apsusc.2007.06.009.

[41] Deng, J.; Zhao, Z.Y. Electronic structure and optical properties of bismuth chalcogenides Bi2Q3 (Q = O, S, Se, Te) by first-principles calculations. Comput. Mater. Sci. **2018**, *142*, 312-319, doi:10.1016/j.commatsci.2017.10.032.

[42] Guimarães, É. V.; Mikhail, H.D.; Silva, A.C.A.; Dantas, N.O.; Silva, R.S. Investigations of structural and optical properties of Bi2−xCrxS3 nanocrystals embedded in host glass. Mater. Lett. **2020**, *265*, doi:10.1016/j.matlet.2020.127430.

[43] Tanabe, Y.; Sugano, S. On the Absorption Spectra of Complex Ions. J. Phys. Soc. Japan **1954**, *9*, 753-766, doi:10.1143/JPSJ.9.753.

[44] Navas, A.S.; Reddy, B.J.; Nieto, F. Spectroscopic study of chromium, iron, OH, fluid and mineral inclusions in uvarovite and fuchsite. *Spectrochim. Acta - Part A Mol. Biomol.* Spectrosc. **2004**, *60*, 2261-2268, doi:10.1016/j.saa.2003.11.024.

[45] Bejenari, I.; Kantser, V.; Balandin, A.A. Thermoelectric properties of electrically gated bismuth telluride nanowires. Phys. Rev. B - Condens. Matter Mater. Phys. **2009**, *81*, 075316, doi:10.1103/PhysRevB.81.075316.

Chapter 9

Antimicrobial Efficacy of Biogenic Silver and Zinc Nanocrystals/Nanoparticles to Combat the Drug Resistance in Human Pathogens

Gulzar Ahmed Rather, Saqib Hassan, Surajit Pal, Mohd Hashim Khan, Heshu Sulaiman Rahman and Johra Khan

Abstract

The emergence of biogenic nanomaterials as novel antimicrobials introduces a new paradigm in human health care. Based on the recent reports of the World Health Organization, infectious diseases pose one of the greatest health challenges. Increased multi-drug resistance prevalence among human pathogens, due to the inefficiency of commercially available antimicrobial drugs in the market is a great threat to humans. The poor solubility, stability and side effects of the antibacterial therapy prompted the researchers to explore new innovative strategies for developing new antimicrobials. Recently, biogenic nanoparticles have proven their effectiveness against multidrug-resistant (MDR) pathogens as an alternative to conventional antibiotics. Biogenic nanoparticles such as silver nanoparticles (AgNPs) and Zinc Oxide nanoparticles (ZnONPs) are easy to produce, biocompatible, provide enhanced uptake and are eco-friendly. Moreover, the capping of the biogenic nanocrystals provides an active surface for interaction with biological components, facilitated by free active surface functional groups to enhance their efficacy and delivery. Inorganic nanocrystals (AgNPs and ZnONPs) are effective both as nano-bactericides and as nanocarriers against sensitive and MDR) pathogens. The present chapter focuses on the utilization of the recent nanosystems to combat drug resistance in human pathogens. Nanomedicine represents a new generation of potiential antimicrobial candidates capable of combating the drug resistance in various pathogenic organisms.

Keywords: Antimicrobials, MDR, AgNPs, ZnONPs, nanocrystals, biogenic nanoparticles

1. Introduction

Nanotechnology is regarded as a new discipline that has a significant influence on human life in several respects with various applications [1]. Current nanoscience research is based on nanoparticle synthesis. Metal nanoparticles are used for a wide range of applications including biosensors, organic marking, cancer therapy, textiles, household and industrial applications etc. Silver nanoparticles AgNPs are mostly used in wound dressings, care of the eye, oral hygiene, biomaterials of bone substitutes, antimicrobial and anti-inflammatory drugs as well as in the coating of catheter products as anti-inflammatory and antimicrobial agents [2]. Silver is a stable and non-hazardous antibiotic agent used for centuries [3]. Most antimicrobials have many disadvantages including low stability, environmental toxicity and the lack of specificity towards the target microorganisms [4]. Few other antimicrobials are extremely irritating and expensive to develop [5].

Silver has the unusual property of binding cellular components that are far larger than nuclear acids [6]. AgNPs may be synthesized employing physical, chemical and biological processes. The majority of the physical and chemical processes of synthesizing nanoparticles have many disadvantages such as low yield, strong reducing agents, energy-intensive mechanisms, uneven particle size and aggregate instability, hazardous waste production, difficulty to scale up and expensive organo-metallic precursors are required [7]. Biological approaches for the synthesis of nanoparticles are regarded as more stable and efficient [8]. For several nanoparticles like gold, silver, platinum and palladium, titanium dioxide, magnetite and cadmium sulphide, the most possible bio-factories are bacteria. Bacteria-mediated AgNPs synthesis is preferred in comparison with other techniques.

Furthermore, bacteria mediated AgNPs are simpler to grow and environment friendly. Both intracellular (biomass) as well as extracellular (cell extracts) synthesis of silver nanoparticles can be performed. Intracellular approaches include the release of synthesized nanoparticles through ultrasonication and additional reactions with specific detergents. It is therefore essential that the AgNPs are synthesized with extracellular methods because of their easy downstream processing that supports large-scale development [7, 9, 10]. There is now a prevalence of multiple tolerance to antibiotics by various clinical infections and pathogens of the urinary tract, caused by excess antibiotics and by an accumulation of antibiotics in the system. This kind of resistance is exhibited by *Staphylococcus* sp., *Streptococcus* sp., *Klebsiella* sp., *Enterococcus* sp., *Proteus* sp., *Pseudomonas* sp. and *E. coli* due to their biofilm-forming potentials [11]. The use of antimicrobial silver nanoparticles will eliminate the multiple-drug resistance, which is a suitable option for antibiotics [12]. Biofilm formation has been regarded as the global barrier in avoiding catheter-related infections in the field of medicine [13]. The conversion of nanoparticles into therapeutic agents, however, involves a detailed knowledge of the physicochemical particularities, results *in vitro* and *in vivo*, biodistribution, pharmacokinetics and pharmacodynamics, apart from the suitable methods of their synthesis [14].

2. Synthesis of nanoparticles

In broader terms, nanoparticles can be synthesized either by (i) Top- down approach, or (ii) Bottom- up approach [15]. Based on the reaction conditions and operation, these two classes can be further categorized as physical, chemical and biological methods [15].

i. Top- down approach

In this method, larger molecules are broken down into smaller units which are then transformed into suitable nanoparticles [15] According to a study, the synthesis of the spherical magnetite nanoparticles that uses natural iron oxide is performed through top- down method which produces particle sizes ranging from 20 nm to 50 nm [15].

ii. Bottom- up approach

This method works in reverse to the top- down approach as nanoparticlessynthesised using this approach are formed from smaller and relatively simpler substances that form clusters and are subsequently converted into desired nanoparticles. This technique is also known as building up approach. Sedimentation and reduction technique fall under this category which include sol gel, green synthesis, spinning and biochemical synthesis [15].

3. Green synthesis of nanoparticles

Green synthesis of nanoparticles refers to the synthesis of nanoparticles through biological routes such as those with the help of microorganisms, enzymes, fungus plants or using various plant products [16, 17] Conventional physical or chemical methods of nanoparticle synthesis often produce byproducts that are hazardous to the environment which is one of the key reasons to opt for a more suitable alternative, that is, the green synthesis or green technology [16]. Other aspects by which green synthesis is more superior than the physical and chemical methods are that they are cost efficient and consume less energy [16].

Bottom- up approach is employed in biological- based synthesis of nanoparticles that requires the use of stabilizing and reducing agents [16]. The process of biologically synthesizing nanoparticles is basically divided into three steps: (i) the choice of a suitable solvent medium used, (ii) the choice of a suitable reducing agent that is eco- friendly and environmentally benign, and (iii) the choice of a non- toxic capping agent that can stabilize the synthesized nanoparticles [16].

Prokaryotes as well as eukaryotes are used in the green synthesis of metallic nanoparticles such as silver, gold, platinum, iron, and metal oxides such as zinc oxide and titanium oxide [17].

3.1 Biological components for green synthesis

Bacteria: Prokarytic bacteria and actinomycetes are widely used in the synthesis of metal and metal oxide nanoparticles as they have the potential to reduce metal ions and therefore, are suitable candidates for the preparation of nanoparticles [18] . The fact that it is relatively easier to manipulate bacteria is a key point in employing them in nanoparticle synthesis [18].

Fungi: Another popular choice for the biological synthesis of metal and metal oxide nanoparticles is fungi as they behave as better biological agents because they have diverse intracellular enzymes [18]. It is also reported that fungi can comparatively synthesize more amounts of nanoparticles than bacteria which could also be because of the fact that fungi have various enzymes/proteins/reducing components on the surface of their cells [18].

Yeast: Saccharomyces cerevisiae has found to be quite effectively employed in the synthesis of silver and gold nanoparticles as reported in numerous studies [18].

Plants: The most simple, efficient, cost effective and feasible method of biosynthesis of metal and metal oxide nanoparticles is using plants and plant extracts as biological agents. Biomolecules such as carbohydrates, proteins and coenzymes extracted from plants are employed to reduce metal salt into nanoparticles [18].

4. Applications of nanoparticles as antimicrobial agents

Nanoparticles have an antimicrobial activity that is capable of overcoming typical resistant mechanisms, including inactivation of the enzymes, reduction of cell permeabilities, modification of target sites/enzymes, and increasing efflux by excessive expression of efflux pumps to escape antimicrobial activity [19, 20]. In addition, NPs combined with antibiotics have a synergistic impact on bacteria, prevent biofilm formation and are used in combating multidrug-resistant organisms [20, 21]. Many features of the nanoparticles provide alternatives to conventional antibiotics. First, with the high volume-surface ratio of nanoparticles, the interaction area with the target species is increased. NP can function as nanoscale molecules that communicate with bacterial cells, regulate penetration of cell membranes and interfere with molecular pathways [22–24]. Secondly, nanoparticles may enhance the inhibitory effects of antibiotics. Saha et al. [25] reported that gold nanoparticles conjugated with ampicillin, streptomycin, or kanamycin could lower the minimum inhibitory concentrations (MICs) of the antibiotic counterparts against both gram-negative and gram-positive bacteria. Gupta et al. [26] have shown a synergistic impact on multi-drug-resistant *Escherichia coli* infections through functionalized gold nanoparticles and fluoroquinolone-based antibiotics.

The complexity of the physicochemical properties, including the scale, form, chemical changes, solvent and environmental factors may influence the antibacterial properties and interfere with the bacteria during the preparation of nanoparticles [27]. Finally, antibiotic and nanoparticle combinations have diverse antimicrobial pathways for overcoming antibiotic resistance [28]. Thus, nanoparticles are regarded as next-generation antibiotics.

Nanoparticles, primarily metallic, have proved effective against gram-positive and gram-negative bacteria in both in vitro and in vivo studies [29]. Even though the antimicrobial mechanisms that are based on the size, shape, zeta-potential, ligands, and material used are not fully understood [28–30] some of the currently accepted mechanisms include (1) disruption of membrane potential and integrity through direct contact with bacteria; (2) activation of the host immune responses; (3) inhibition of biofilm formation; (4) generation of reactive oxygen species (ROS); and (5) induction of intracellular effects to inhibit RNA and protein synthesis [21, 27]. Nanoparticle coatings on implantable devices, urinary catheters, wound dressings, bone cement, or dental implants/materials can function as nanoparticle-based antibiotic delivery systems [31, 32]. Furthermore, nanoparticles can be used as vectors to transfer and deliver higher doses of drugs to infected sites [21]. Thus, the combination of NPs and antimicrobial agents may be beneficial in fighting the ongoing crisis of antimicrobial resistance [20]. The combination of nanoparticles and antimicrobials could thus help to combat the current anti-microbial resistance crisis [20].

Nanoparticles with antimicrobial activity that combat *E. faecium, S. aureus, K. pneumoniae, A. baumannii, P. aeruginosa, and Enterobacter species* [24, 27, 33, 34] include the nanoparticles containing silver (Ag), gold (Au), zinc (Zn), copper (Cu), Titanium (Ti), magnesium (Mg), nickel (Ni), cerium (Ce), selenium (Se), aluminum (Al), cadmium (Cd), yttrium (Y), palladium (Pd), or superparamagnetic iron [24]. It has been reported that among various metallic nanoparticles and their oxides, silver or its ionic form is most toxic to bacteria [35]. Silver nanoparticles (AgNPs) find numerous uses as they possess several mechanisms of antibacterial activity [36], high biocompatibility, and functionalized potential and are easily detectable [37]. Even though AgNPs are difficult to functionalize with biomolecules and antibiotics, Ag–gold (Au) alloys provide an alternative, since they combine the antimicrobial effects of Ag with the effectiveness of functionalization and the stability of Au in the form of bimetallic NPs [20]. Furthermore, Ag–Au NPs functionalized with tetracycline have been shown to have a synergetic effect, which is attributed to the generation of ROS [38].

Antibiotic resistance is often associated with biofilm formation and an active quorum sensing machinery within a bacterial cell. Biofilm plays a crucial role by evading the action of antibiotics that hinder their entry in the cell membrane surface as a result rapid dissemination of resistance may occur. Almost all bacteria posses biofilm and it may remain in an inert state by modulating the epigenetic changes within its genome allowing them to develop adaptive resistance as a result. Exopolysaccharide capsular fractions, extracellular polymeric substances (EPS), also provide structural integrity that works as a barrier, restrict the antibiotic action by modulating the cellular integrity, further promote antimicrobial resistance. Several evidences have shown the capability of nanoparticle to disrupt the cell membrane, hinder biofilm formation and dispute the active quorum-sensing systems. The design of biologically active nanoparticle has to be made in such a way that it must circumvent the barrier function to elicit their activity. The most predominant and promising strategy could be to interfere with the cellular signaling mediated by quorum sensing molecules such as Acyl homoserine lactone (AHLs) for Gram-negative bacteria and peptides of Gram-positive bacteria. (**Figure 1**).

The mechanism of the nanoparticle often dealt with the synthesis of N-acetylated homoserine lactonase proteins (AiiA) capable of modulating signals by preventing the stimuli from reaching its cognate receptor site, as a result, the intracellular signaling affects the communication. The activity of quorum quenching could be enhanced by coupling the drug moieties into a nanoformulation. Enormous alternatives are available, capable enough to elicit its activity by employing dendrimers, chitosan and liposome coating to give effective drug delivery. Siddhardha et al. [39], have recently evaluated the anti-biofilm property of the chrysin-loaded chitosan nanoparticles and their role in combating infections caused by *S.aureus*. Chrysin, a plant flavone constituent of *Orocylumineicum vent* is well-established for its biological properties. However, its therapeutic attributes have

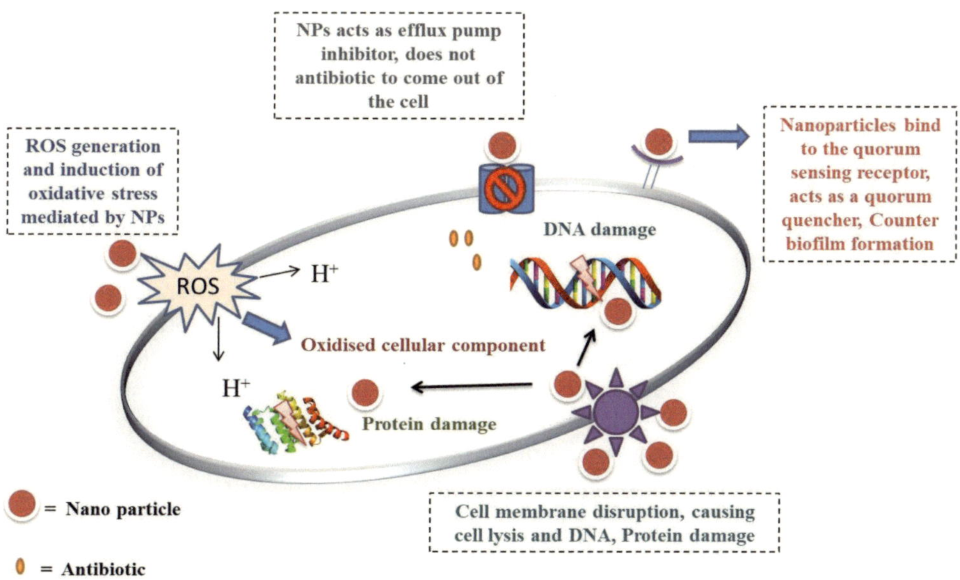

Figure 1.
Mechanism of antimicrobial activity by nanoparticles.

not been fully deciphered due to its poor solubility and less bioavailability. In their study, chrysin has been encapsulated into chitosan derived nanoparticles using TPP as a linker.

Nanoparticles were further characterized and assessed for their ability as an anti-biofilm compound against *Staphylococcus aureus*. In sub-Minimum Inhibitory Concentration, (MICs) these nanoparticles exhibited increased anti-biofilm efficacy against *S. aureus* as compared to its counterparts, chrysin and chitosan, remarkably reduced in the cell surface hydrophobicity index and exopolysaccharide (EPS) production indicated by the inhibitory effect of nanoparticles on the primary stages of biofilm formation. However, growth curve analysis further showed that at a sub-MIC point nanoparticles did not exert any bactericidal effect against *S. aureus*. During the last few years, various nano-mediated delivery systems such as poly(lactic-co-glycolic acid) (PLGA), fusogenic liposome molecules, solid lipid nanoparticles, also known as SLNs, are physiological lipids dispersed in aqueous surfactant and lipid-polymer hybrid have proven to be promising vehicles to act as a nanocarrier. Commercially available Protein, and lipid conjugated polymeric substance viz. Intron A and AmBiosome have become the most predominant nano platform to evaluate their efficacy in terms of countering the biofilm formation within pathogenic bacteria. In one of the recent studies, sodium lauryl sulfate (SLSs) have been shown to act as anti quorum sensing agents (Quorum quencher) by downregulating the pyocyanin production in *P. aeruginosa* strain.

5. Silver nanoparticles and their antibacterial efficacy

Silver nanoparticles have been identified as an excellent antibacterial agent with potential medical applications. Green nanoparticle synthesis has emerged as a viable alternative to overcome the limitations of traditional methods of nanoparticle synthesis [40]. Green synthesis of silver nanoparticles (AgNPs) employs plant constituents such as sugars, fats, flavonoids, alkaloids, polyphenols have become an alternative to the rational chemical formulation. To date, green synthesis of AgNPs is carried out by using fruit extract of *Emblica officinalis*, leaves extract of *Citrus limon*, *Azadirachta indica*, *Coffea Arabica*, *Acalypha indica*, *Camellia sinensis*, a root extract of *Phoenix dactylifera*, *Morinda citrifolia*, inflorescence extract of *Mangifera indica* etc. [41]. Phytochemicals exhibit various antimicrobial activities and they act as reducing or stabilizing agents in plants and the co-occurrence of these natural compounds shape the biogenic AgNPs morphology.

Paradoxically Mechanism of action of AgNPs for eliciting antibacterial efficacy is a controversial topic. Several research papers have shared certain assumptions such as Ag ions get released by the AgNPs and submerged around the cell surface affecting the penetration barrier of the cell wall or cell surface by dismantling the cellular permeability. It has been reported that AgNPs can affect DNA replication while interacting with thiol moieties of protein molecule initiate protein deactivation with the concomitant liberation of reactive oxygen species (ROS) [42].

Other hypotheses suggest that the receptor musking capabilities of AgNPs due to their size which might restrict the biofilm formation in *E.coli* and *S.aureus* by disputing the quorum sensing machinery. The dissolution status of AgNPs strongly affects their antibacterial efficacy depends on synthetic and intrinsic characteristics of surrounding media. The coexistence of particle size and morphology on the liberation of Ag + described, by the Ostwald–Freundlich equation. The presence of smaller AgNPs with quasi-spherical shape are more prone to Ag + release, due to their greater surface ratio. This quality further explains the lower silver content of aggregated nanoparticles, which is relative to isolated nanoparticles. Certain capping agents are used to modulate the AgNPs surface, which in turn, may cause their dissolution behavior. Sometimes, the surrounding media itself can influence the release of Ag + ions. The co-occurrence of inorganic or organic components in a growth medium can affect the solvation potential of silver nanoparticles by augmenting with nascent AgNPs or coupling with silver ions. Studies have also shown that AgNPs release Ag + faster in acidic condition than they do in neutral solution [43].

Certain Gram-negative bacteria are more susceptible to AgNPs because of their hallmark cellular architecture, which is the presence of an outer membrane coating over the polymeric peptidoglycan layer. On the other hand, the cell wall of gram-negative bacteria is much narrower than that of gram-positive bacteria, hence, the thick cell wall might reduce the penetration of NPs into cells. Moreover, the antibacterial effects of AgNPs on gram-negative and gram-positive cell emphasize that uptake of AgNPs is crucial to elicit its antibacterial effect.

Silver nanoparticles, smaller than 10 nm can directly alter cell permeability, enter bacterial cells and cause cell damage by employing the abovementioned strategies. Most of the noxious biofilm formers, rapidly develop exopolymeric substances which protect bacteria from both Ag + and NPs activity by hindering their transport. Researchers have observed that 100% reduction or mortality of bacteria did not occur in the biofilm state, whereas, AgNPs with the same concentration, enabled their effect by killing all planktonic cells. However, the polymeric biofilm hinders the use AgNPs, due to its complicated cellular architecture. Diffusion coefficients of AgNPs are generally related to size, shape and physicochemical characteristics, which governs their mobility and bioavailability in biofilm structure. The diffusion coefficient decreases with increasing molar mass make it harder for larger AgNPs to counter biofilm formation. Inter and Intracellular transport through bacterial biofilm can be impacted for particles >50 nm. Chemical constituents of NPs can also arouse adsorption augmentation of AgNPs in the biofilm, by interfering with their diffusion profile. Lastly, electrostatic intricacy among bacteria and AgNPs could influence charged nanoparticles' penetration through biofilm [44].

Silver is being considered to be a potent antimicrobial agent. Silver nanoparticles (AgNPs) are the most promising inorganic NPs employed to treat a variety of bacterial infections. Synthesis of AgNPs could be achieved by green syntheses such as using plant or microbial extracts. AgNPs mediated cell death, causing cell wall disruption has become one of the most striking opportunities for the antimicrobial researcher to evaluate its efficacy *in vitro.* AgNPs are capable of producing their activity inactivating the respiratory chain followed by ROS production with a process of oxidation of the bacterial cellular constituents. The cell permeability increases upon treating with AgNPs, cause depolarization of the cell wall. AgNPs have been found to show synergized effect with different antibiotics that resulted in better efficacy. Lara et al. [45] have demonstrated the efficacy of AgNPs against multidrug-resistant nosocomial pathogenic bacteria such as *P.aeruginosa*, ampicillin-resistant *E.coli* O15:H7 and MDR *Streptococci* strain. It has been reported that Ag-bearing nanocrystals are sensitive against a variety of Gram-positive and Gram-negative bacteria such as Vancomycin-resistant *Enterococci*, *S. mearcesens*, counter the biofilm formation in MTP assays.

Antimicrobial peptides (AMPs) prove to be one of the key elements for the defense strategy against various high-density bacterial infections. Unfortunately, AMPs are sparingly soluble and have possess poor enzymatic stability, often get churned into pieces before they cross the biological barrier due to their low permeability. Immobilization of AMPs onto NPs can be an exciting way for drug delivery. Polymyxin B is the most used AMP, as it has a good antibacterial activity, it modulates the LPS in the outer membrane present within Gram-negative cells. It has been demonstrated recently that AgNPs coupled with polymyxin-B removed endotoxin contamination from the surgical solutions and downregulated the biofilm formation on the blades used for surgical procedures.

6. Zinc oxide nanoparticles and their antibacterial efficacy

Zinc oxide nanoparticles are utilized in food as additives, supplements, containers, and packaging; in the energy sector as fuels and catalysts; in consumer electronics, semiconductors, and air filters; in biomedical engineering; and in drinking water.

Plant constituents such as *Cassia tora*, *Calotropis gigantean*, *Hibiscus sp*, *Corymbia citriodora* etc. have been used for green synthesis of ZnNPs [46, 47]. These plants for example, *Catharanthus roseus* produce secondary metabolites, containing more than 200 terpenoid based indole alkaloids, notable for their antimicrobial and anticancer activity [48].

Recently, the antibacterial efficacy of Zn NPs was assessed for *Staphylococcus aureus* MTCC 9760, *Pseudmonas aeruginosa* MTCC 424, *Streptococcus pyogenes* MTCC 1926, *Escherichia coli* MTCC 40, *Bacillus cereus* MTCC 430 and *Proteus mirabilis* MTCC 3310. The green newly synthesized Zn NPs exhibited promising antibacterial activity against both Gram-positive and Gram-negative bacteria. These results elucidated a rapid, cost-effective, environment friendly and convenient method for ZnO NPs synthesis, which could be used as a potential antimicrobial agent against drug-resistant microbes [49].

Zinc nanoparticles minimize the use of toxic substances in NPs fabrication and modulate the antibacterial efficacy and biomedical applications. Several studies have reported the efficacy of ZnO NPs against various pathogenic bacteria. It has been observed that the shape and size of ZnO NPs dependent on several physicochemical characteristics such as reaction kinetics, annealing temperature and pH. ZnONPs exhibit a strong antimicrobial efficacy for *E. coli* and *S. aureus* without UV illumination at NP concentration of 0.025 mg mL^{-1} after 8 h of incubation [50].

Green synthesis of Zn NPs could be initiated by rapid mixing of the aqueous solution of 0.01 M zinc acetate dehydrate with above mentioned any of the leaf extracts. This process is further followed by constant stirring till the appearance of white suspension. The pH needs to be adjusted till the ZnO NPs precipitate is completely dissolved. Finally, the spectra exhibites an absorption band with a resolution of 1.0 nm between 350 and 500 nm, which confirms the formation of ZnO NPs. Physicochemical parameters regulate the green synthesis of inorganic metal NPs. Hence, several physicochemical parameters are optimized such as pH, temperature; reaction time and concentration of the metal ions [51].

Zinc oxide NPs are characterized for their bactericidal activity by affecting planktonic cell growth and their proliferation. Like AgNPs the mechanism is to produce ROS with an adjunct to its Zn^{2+} release in a liquid medium, which hinders the enzymatic activity and normal physiological process within a healthy bacterial cell. The stability of these oxides is achieved by their high density, high melting point, high thermal conductance and conductivity. Besides ZnO, rare-earth NPs have become one of the alternative treatment choices and their oxide forms differ in chemistry from the main group elements as well as transition metals because the 4*f* orbitals of these elements are buried deep within the atom and the 4*d* and 5*p* electrons shield the 4*f* orbitals from the atom's environment and thus owing to the unique characteristics of these elements in terms of catalytic, magnetic and electronic properties which can be exploited by preparing nanoparticles from their oxide forms. Sarwar and his coworkers [52] have recently demonstrated that the nanoformulation of ZnO has caused significant oxidative stress to *Vibrio cholera*, with a concominant downstream process of DNA degradation, protein leakage, membrane depolarization and fluidity as a result.

Moreover, ZnO NPs impregnated with antibiotics have shown better efficacy to act as an antimicrobial agent against almost every noxious biofilm formers under the group of ESKAPE pathogens viz. *S. aureus*, *Proteus* sp., *Acinetobacter* sp., *P. aeruginosa*, and *E. coli*, whereas the antibiotic alone found to be resistant towards them, showed a synergistic effect under these circumstances.

7. Nanoparticle based alloys

Nanoparticle alloys made by the combination of Ag and Au may.be used to enhance the effects of a drug. Even tough they have antimicrobial effects when used alone, once they are used in combination their effects get enhanced [30, 53]. It has been reported that AgNPs face difficulties in functionalizing with biomolecules and drugs, to overcome this limitation in monometallic NPs, alloy/bimetallic NPs have been introduced which possess improved electronic, optical and catalytic properties [54, 55]. AuNPs are considered to be good vectors for the delivery of pharmacologic compounds. The combinatorial effect of nanoparticle alloys increases efficacy- in biological media, gold enhances the functionalization improves the stability along with the antimicrobial activity of silver [56, 57]. Fakhri et al. [38] found that antibiotic (tetracycline) along with the bimetallic nanoparticles (Au-AgNPs.) showed a synergetic effect, with greater antimicrobial activity. Baker et al. [58] in their study have reported antimicrobial activity of bimetallic AgAuNPs against *Pseudomonas veronii* strain AS41G inhibiting *Annona squamosa L.* They also reported their synergistic effect with standard antibiotics against the bacitracin resistant strains of *Bacillus subtilis, E. coli*, and *K. pneumoniae* [58]. Zhao et al. [59] have signified the antibacterial activity of bimetallic NPs like AuPtNPs against drug resistant bacteria.

8. Synergistic effects of NPs with antibiotics

For overcoming antibiotic resistance, NPs can be customized and packaged with various antimicrobial agents. NPs use several mechanisms to combat bacterial resistance. If NPs are used in combination with antibiotics, there is very less chance that bacteria can develop resistance [60, 61]. Thus, NPs in combination with antibiotics can be a promising strategy to overcome bacterial resistance. Additionally, NPs are effective antimicrobial delivery systems, and may reduce the dosage and toxicity of antibiotics [24]. Very low concentrations of AgNPs and antibiotics have been synergistically found to inhibit various pathogens such as *S. aureus*, *P. aeruginosa*, *A. baumannii* strains etc. [62–64]. Similarly, synergistic effects Ag, Au, and ZnO NPs along with antibiotics have been found to be effective against *S. aureus*, *E. faecium*, *E. coli*, *A. baumannii*, and *P. aeruginosa* [24]. The efficacy of antibiotics along with functionalized NPs may help in reversal of antimicrobial resistance and may also enhance the antimicrobial effects of various antibiotics [24].

9. Nanotoxicity

Although there is an immense progress in the field of nanotechnology, but the consequent health effects that are related to the exposure to nanoparticles remain largely unexplored. Researchers have started to characterize the risks associated with nanoparticle exposure. [65]. Impact of NPs on beneficial bacteria in humans should be of high concern [66, 67]. It has been reported that NPs can cause hemeolysis thus, may impair blood coagulation [68]. The precise mechanism of toxic effects of NPs is not so clear, but it has been observed that nanoparticles larger in size pose a greater threat on human health. In most of the invitro studies conducted on AgNPs, it has been observed that AgNPs are more toxic for cell lines [69, 70]. Deposition of AgNPs has been observed in many organs such as liver, lungs, spleen which has been linked to organ damage [24]. In blood and urine of burn patients elevated levels of Ag have been reported, this may be due to leaching from Ag wound dressing (Acticoat®),which is composed of Ag nanocrystals, into the blood stream [67, 71].

Neurotoxicity has been associated with Al2O3 NPs which interact with cellular components [33]. CuONPs have been observed to induce oxidative damage. CuO NPs induce hepatotoxicity and nephrotoxicity through the generation of free radical-mediated oxidative stress [20, 72, 73]. DNA damage has been associated with ZnO or TiO2 NPs thus, limiting their use [24]. Hagens et al. [74] have reported that NPs that are administered intravenously could get accumulated in bone marrow, colon, lung, liver, spleen as well as lymphatic system. Inhalation of NPs has been associated with cytotoxicity in the lungs [75].

In various *in vivo* studies, it has been reported that there is lethal toxicity associated with the use of NPs [29, 76–78]. Nevertheless, the assessment of toxicity at the cellular level as well as mode of administration is crucial for clinical use of NPs [67].

10. Conclusion

Silver and Zinc NPs have the ability to interact with and influence the growth of a variety of microorganisms. Therefore, Ag and Zn NPs could be employed as a broad-spectrum antibiotic agents to combat bacterial infections. Plant extracts are used to make Ag and Zn NPs, which is cost-effective and eco-friendly. The antibacterial action of Ag and Zn NPs appears to be attributed to their structure, ultrasmall size, and increased surface area, which allows them to damage and pass the bacterial membrane that is responsible for intracellular damage. It has been described that NPs in combination with antibiotics can be a promising strategy to overcome bacterial resistance. Synthesis of Ag and Zn NPs via green-synthesis methods and mechanisms of action against bacteria have been elucidated. We have also shed light on toxicity associated with the use of NPs. Ultimately, it can be conluded that both Ag and Zn sNP have broad-spectrum antibiotic activity against bacteria, making them prospective therapeutic agents for bacterial infections and multidrug-resistant pathogens.

Author details

Gulzar Ahmed Rather[1]*, Saqib Hassan[2,3], Surajit Pal[2], Mohd Hashim Khan[4], Heshu Sulaiman Rahman[5] and Johra Khan[6]

1 Department of Biomedical Engineering, Sathyabama Institute of Science and Technology, Deemed to be University, Tamil Nadu, India

2 Department of Microbiology, School of Life Sciences, Pondicherry University, Puducherry, India

3 Division of Non-Communicable Diseases, Indian Council of Medical Research (ICMR), New Delhi, India

4 School of Life Sciences, B.S. Abdur Rahman Crescent Institute of Science and Technology, Tamil Nadu, India

5 Department of Physiology, College of Medicine, University of Sulaimani Kurdistan Region, Iraq

6 Department of Medical Laboratory Sciences, College of Applied Medical Sciences, Majmah University, Kingdom of Saudi Arabia

*Address all correspondence to: rathergulzar5@gmail.com

IntechOpen

© 2021 The Author(s). Licensee IntechOpen. This chapter is distributed under the terms of the Creative Commons Attribution License (http://creativecommons.org/licenses/by/3.0), which permits unrestricted use, distribution, and reproduction in any medium, provided the original work is properly cited.

References

[1] Nikalje AP. Nanotechnology and its applications in medicine. Medicinal Chemistry. 2015; 5:081-089.

[2] Salata OV. Applications of nanoparticles in biology and medicine. Journal of Nanobiotechnology. 2004; 2:1-6.

[3] Lansdown AB. Silver in health care: antimicrobial effects and safety in use. Biofunctional textiles and the skin. 2006; 33:17-34.

[4] Butola BS, Mohammad F. Silver nanomaterials as future colorants and potential antimicrobial agents for natural and synthetic textile materials. RSC Advances. 2016;6: 44232-44247.

[5] Sondi I, Salopek-Sondi B. Silver nanoparticles as antimicrobial agent: a case study on *E. coli* as a model for Gram-negative bacteria. Journal of Colloid and Interface Science. 2004; 275:177-182.

[6] Clement JL, Jarrett PS. Antibacterial silver. Metal-based Drugs. 1994; 1:467-482.

[7] Iravani S. Bacteria in nanoparticle synthesis: current status and future prospects. International scholarly research notices. 2014.

[8] Pantidos N, Horsfall LE. Biological synthesis of metallic nanoparticles by bacteria, fungi and plants. Journal of Nanomedicine & Nanotechnology. 2014; 5:1.

[9] Deepak V, Kalishwaralal K, Pandian SR, Gurunathan S. An insight into the bacterial biogenesis of silver nanoparticles, industrial production and scale-up. InMetal nanoparticles in Microbiology 2011;17-35. Springer, Berlin, Heidelberg.

[10] Das VL, Thomas R, Varghese RT, Soniya EV, Mathew J, Radhakrishnan EK. Extracellular synthesis of silver nanoparticles by the *Bacillus* strain CS 11 isolated from industrialized area. 3 Biotech. 2014; 4:121-126.

[11] Flores-Mireles AL, Walker JN, Caparon M, Hultgren SJ. Urinary tract infections: epidemiology, mechanisms of infection and treatment options. Nature Reviews Microbiology. 2015;13:269-284.

[12] Salomoni R, Léo P, Montemor AF, Rinaldi BG, Rodrigues MF. Antibacterial effect of silver nanoparticles in Pseudomonas aeruginosa. Nanotechnology, Science and Applications. 2017; 10:115.

[13] Batoni G, Maisetta G, Esin S. Antimicrobial peptides and their interaction with biofilms of medically relevant bacteria. Biochimica et Biophysica Acta (BBA)-Biomembranes. 2016;1858:1044-1060.

[14] Burduşel AC, Gherasim O, Grumezescu AM, Mogoant L, Ficai A, Andronescu, E. Biomedical applications of silver nanoparticles: an up-to-date overview, Nanomaterials (Basel), 2018.

[15] Khan I, Saeed K, Khan I. Nanoparticles: Properties, applications and toxicities. Arabian Journal of Chemistry. 2019;12:908-931.

[16] Patra JK, Baek KH. Green nanobiotechnology: factors affecting synthesis and characterization techniques. Journal of Nanomaterials. 2014.

[17] Hasan S. A review on nanoparticles: their synthesis and types. Research Journal of Recent Sciences. 2015; 2277:2502.

[18] Singh J, Dutta T, Kim KH, Rawat M, Samddar P, Kumar P. 'Green'synthesis of

metals and their oxide nanoparticles: applications for environmental remediation. Journal of Nanobio technology. 2018; 16:1-24.

[19] Mulvey MR, Simor AE. Antimicrobial resistance in hospitals: how concerned should we be?. Cmaj. 2009; 180:408-415.

[20] Baptista PV, McCusker MP, Carvalho A, Ferreira DA, Mohan NM, Martins M, Fernandes AR. Nano-strategies to fight multidrug resistant bacteria—"A Battle of the Titans". Frontiers in icrobiology. 2018; 9:1441.

[21] Pelgrift, RY, and Friedman AJ. Nanotechnology as a therapeutic tool to combat microbial resistance. Advanced Drug Delivery Reviews.2013; 65:1803-1815. doi:10.1016/j.addr.2013.07.011

[22] Rai MK, Deshmukh SD, Ingle AP, Gade AK. Silver nanoparticles: the powerful nanoweapon against multidrug-resistant bacteria. Journal of Applied Microbiology. 2012;112:841-852.

[23] Durán N, Durán M, de Jesus MB, Fávaro WJ, Nakazato G, Seabra AB. Antimicrobial activity mechanisms of silver nanoparticles: An overview. Nanomedicine. 2016;12:789-799.

[24] Hemeg HA. Nanomaterials for alternative antibacterial therapy. International journal of nanomedicine. 2017;12:8211.

[25] Saha B, Bhattacharya J, Mukherjee A, Ghosh A, Santra C, Dasgupta AK, Karmakar P. In vitro structural and functional evaluation of gold nanoparticles conjugated antibiotics. Nanoscale Research Letters. 2007;2:614-622.

[26] Gupta A, Saleh NM, Das R, Landis RF, Bigdeli A, Motamedchaboki K, Campos AR, Pomeroy K, Mahmoudi M, Rotello VM. Synergistic antimicrobial therapy using nanoparticles and antibiotics for the treatment of multidrug-resistant bacterial infection. Nano Futures. 2017;1:015004.

[27] Beyth N, Houri-Haddad Y, Domb A, Khan W, Hazan R. Alternative antimicrobial approach: nano-antimicrobial materials. Evid. Based. Complementary Alternative Medicine. 2015;246012. doi:10.1155/2015/246012

[28] Huh AJ, and Kwon YJ. "Nanoantibiotics": a new paradigm for treating infectious diseases using nanomaterials in the antibiotics resistant era. Journal of Controlled Release. 2011; 156, 128-145. doi:10.1016/j.jconrel.2011.07.002

[29] Zazo H, Colino CI, Lanao JM. Current applications of nanoparticles in infectious diseases. Journal of Controlled Release. 2016 Feb 28;224:86-102.

[30] Singh R, Smitha MS, Singh SP. The role of nanotechnology in combating multi-drug resistant bacteria. Journal of Nanoscience and Nanotechnology. 2014;14:4745-4756.

[31] Wang L, Hu C, Shao L. The antimicrobial activity of nanoparticles: present situation and prospects for the future. International Journal of Nanomedicine. 2017;12:1227.

[32] Divya M, Kiran GS, Hassan S, Selvin J. Biogenic synthesis and effect of silver nanoparticles (AgNPs) to combat catheter-related urinary tract infections. Biocatalysis and agricultural biotechnology. 2019;18:101037.

[33] Ansari MA, Khan HM, Khan AA, Cameotra SS, Saquib Q, Musarrat J. Interaction of A l2 O 3 nanoparticles with E scherichia coli and their cell envelope biomolecules. Journal of Applied Microbiology. 2014;116: 772-783.

[34] Dizaj SM, Lotfipour F, Barzegar-Jalali M, Zarrintan MH, Adibkia K. Antimicrobial activity of the metals and metal oxide nanoparticles. Materials Science and Engineering: C. 2014; 44:278-284.

[35] Seil JT, Webster TJ. Antimicrobial applications of nanotechnology: methods and literature. International Journal of Nanomedicine. 2012;7:2767.

[36] Cheng G, Dai M, Ahmed S, Hao H, Wang X, Yuan Z. Antimicrobial drugs in fighting against antimicrobial resistance. Frontiers in Microbiology. 2016;8:470.

[37] Baranwal A, Srivastava A, Kumar P, Bajpai VK, Maurya PK, Chandra P. Prospects of nanostructure materials and their composites as antimicrobial agents. Frontiers in Microbiology. 2018;9:422.

[38] Fakhri A, Tahami S, Naji M. Synthesis and characterization of core-shell bimetallic nanoparticles for synergistic antimicrobial effect studies in combination with doxycycline on burn specific pathogens. Journal of Photochemistry and Photobiology B: Biology. 2017;169:21-26.

[39] Siddhardha B, Pandey U, Kaviyarasu K, Pala R, Syed A, Bahkali AH, Elgorban AM. Chrysin-Loaded Chitosan Nanoparticles Potentiates Antibiofilm Activity against *Staphylococcus aureus*. Pathogens. 2020;9:115.

[40] Rajeshkumar S, Bharath LV, Geetha R. Broad spectrum antibacterial silver nanoparticle green synthesis: Characterization, and mechanism of action. InGreen synthesis, characterization and applications of nanoparticles 2019;1:429-444.

[41] Tungmunnithum D, Thongboonyou A, Pholboon A, Yangsabai A. Flavonoids and other phenolic compounds from medicinal plants for pharmaceutical and medical aspects: An overview. Medicines. 2018;5:93.

[42] Yun'an Qing LC, Li R, Liu G, Zhang Y, Tang X, Wang J, Liu H, Qin Y. Potential antibacterial mechanism of silver nanoparticles and the optimization of orthopedic implants by advanced modification technologies. International Journal of Nanomedicine. 2018;13:3311.

[43] Yin IX, Zhang J, Zhao IS, Mei ML, Li Q, Chu CH. The antibacterial mechanism of silver nanoparticles and its application in dentistry. International Journal of Nanomedicine. 2020;15:2555.

[44] Liao C, Li Y, Tjong SC. Bactericidal and cytotoxic properties of silver nanoparticles. International Journal of Molecular Sciences. 2019;20:449.

[45] Lara HH, Ayala-Núnez NV, Turrent LD, Padilla CR. Bactericidal effect of silver nanoparticles against multidrug-resistant bacteria. World Journal of Microbiology and Biotechnology. 2010;26:615-621.

[46] Iavicoli I, Leso V, Beezhold DH, Shvedova AA. Nanotechnology in agriculture: Opportunities, toxicological implications, and occupational risks. Toxicology and Applied Pharmacology. 2017;329:96-111.

[47] Schmalz G, Hickel R, Van Landuyt KL, Reichl FX. Scientific update on nanoparticles in dentistry. International Dental Journal. 2018;68:299-305.

[48] Tiwari V, Mishra N, Gadani K, Solanki PS, Shah NA, Tiwari M. Mechanism of anti-bacterial activity of zinc oxide nanoparticle against carbapenem-resistant *Acinetobacter baumannii*. Frontiers in Microbiology. 2018;9:1218.

[49] Gupta M, Tomar RS, Kaushik S, Mishra RK, Sharma D. Effective antimicrobial activity of green ZnO nano particles of Catharanthus roseus. Frontiers in microbiology. 2018;9:2030.

[50] Thi TU, Nguyen TT, Thi YD, Thi KH, Phan BT, Pham KN. Green synthesis of ZnO nanoparticles using orange fruit peel extract for antibacterial activities. RSC Advances. 2020;10:23899-23907.

[51] FAL HN, Farzaneh F. Synthesis of ZnO nanocrystals with hexagonal (Wurtzite) structure in water using microwave irradiation. 2006;231-234.

[52] Sarwar A, Katas H, Samsudin SN, Zin NM. Regioselective sequential modification of chitosan via azide-alkyne click reaction: synthesis, characterization, and antimicrobial activity of chitosan derivatives and nanoparticles. PLoS One. 2015;10:e0123084.

[53] Arvizo R, Bhattacharya R, Mukherjee P. Gold nanoparticles: opportunities and challenges in nanomedicine. Expert Opinion on Drug Delivery. 2010;7:753-763.

[54] Cho KH, Park JE, Osaka T, Park SG. Silver nanoparticles in cosmetics. Electrochimica Acta. 2005;51: 956-960.

[55] Shah A, Khan SB, Asiri AM, Hussain H, Han C, Qureshi R, Ashiq MN, Zia MA, Ishaq M, Kraatz HB. Synthesis, characterization, and application of Au–Ag alloy nanoparticles for the sensing of an environmental toxin, pyrene. Journal of Applied Electrochemistry. 2015;45:463-472.

[56] Doria G, Larguinho M, Dias JT, Pereira E, Franco R, Baptista PV. Gold–silver-alloy nanoprobes for one-pot multiplex DNA detection. Nanotechnology. 2010;21:255101.

[57] Dos Santos MM, Queiroz MJ, Baptista PV. Enhancement of antibiotic effect via gold: silver-alloy nanoparticles. Journal of Nanoparticle Research. 2012;14:1-8.

[58] Baker S, Pasha A, Satish S. Biogenic nanoparticles bearing antibacterial activity and their synergistic effect with broad spectrum antibiotics: emerging strategy to combat drug resistant pathogens. Saudi Pharmaceutical Journal. 2017;25:44-51.

[59] Zhao Y, Ye C, Liu W, Chen R, Jiang X. Tuning the composition of AuPt bimetallic nanoparticles for antibacterial application. Angewandte Chemie International Edition. 2014; 53:8127-8131.

[60] Fischbach MA. Combination therapies for combating antimicrobial resistance. Current Opinion in Microbiology. 2011;14:519-523.

[61] Zhao Y, Jiang X. Multiple strategies to activate gold nanoparticles as antibiotics. Nanoscale. 2013;5:8340-8350.

[62] Naqvi SZ, Kiran U, Ali MI, Jamal A, Hameed A, Ahmed S, Ali N. Combined efficacy of biologically synthesized silver nanoparticles and different antibiotics against multidrug-resistant bacteria. International Journal of Nanomedicine. 2013;8:3187.

[63] Panáček A, Smékalová M, Kilianová M, Prucek R, Bogdanová K, Večeřová R, Kolář M, Havrdová M, Płaza GA, Chojniak J, Zbořil R. Strong and nonspecific synergistic antibacterial efficiency of antibiotics combined with silver nanoparticles at very low concentrations showing no cytotoxic effect. Molecules. 2016; 21:26.

[64] Scandorieiro S, de Camargo LC, Lancheros CA, Yamada-Ogatta SF, Nakamura CV, de Oliveira AG, Andrade CG, Duran N, Nakazato G,

Kobayashi RK. Synergistic and additive effect of oregano essential oil and biological silver nanoparticles against multidrug-resistant bacterial strains. Frontiers in Microbiology. 2016;7:760.

[65] Gupta I, Duran N, Rai M. Nano-silver toxicity: emerging concerns and consequences in human health. InNano-Antimicrobials, Springer; 2012. p. 525-548.

[66] Zhang L, Pornpattananangkul D, Hu CM, Huang CM. Development of nanoparticles for antimicrobial drug delivery. Current Medicinal Chemistry. 2010;17:585-594.

[67] Khan ST, Musarrat J, Al-Khedhairy AA. Countering drug resistance, infectious diseases, and sepsis using metal and metal oxides nanoparticles: current status. Colloids and Surfaces B: Biointerfaces. 2016;146:70-83.

[68] Kandi V, Kandi S. Antimicrobial properties of nanomolecules: potential candidates as antibiotics in the era of multi-drug resistance. Epidemiol Health. 2015;37:e2015020. doi: 10.4178/epih/e2015020.

[69] Bondarenko O, Juganson K, Ivask A, Kasemets K, Mortimer M, Kahru A. Toxicity of Ag, CuO and ZnO nanoparticles to selected environmentally relevant test organisms and mammalian cells in vitro: a critical review. Archives of Toxicology. 2013; 87:1181-1200.

[70] Ivask A, Juganson K, Bondarenko O, Mortimer M, Aruoja V, Kasemets K, Blinova I, Heinlaan M, Slaveykova V, Kahru A. Mechanisms of toxic action of Ag, ZnO and CuO nanoparticles to selected ecotoxicological test organisms and mammalian cells in vitro: a comparative review. Nanotoxicology. 2014; 8:57-71.

[71] Vlachou E, Chipp E, Shale E, Wilson YT, Papini R, Moiemen NS. The safety of nanocrystalline silver dressings on burns: a study of systemic silver absorption. Burns. 2007;33:979-985.

[72] De Jong WH, and Borm PJ. Drug delivery and nanoparticles:applications and hazards. International Journal of Nanomedicine 2008;3:133-149. doi: 10.2147/IJN.S596.

[73] Lei R, Wu C, Yang B, Ma H, Shi C, Wang Q, Wang Q, Yuan Y, Liao M. Integrated metabolomic analysis of the nano-sized copper particle-induced hepatotoxicity and nephrotoxicity in rats: a rapid in vivo screening method for nanotoxicity. Toxicology and Applied Pharmacology. 2008; 232:292-301.

[74] Hagens WI, Oomen AG, de Jong WH, Cassee FR, Sips AJ. What do we (need to) know about the kinetic properties of nanoparticles in the body?. Regulatory Toxicology and Pharmacology. 2007;49:217-229.

[75] Leucuta SE. Systemic and biophase bioavailability and pharmacokinetics of nanoparticulate drug delivery systems. Current Drug Delivery.2013;10; 208-240.

[76] Pfurtscheller K, Petnehazy T, Goessler W, Bubalo V, Kamolz LP, Trop M. Transdermal uptake and organ distribution of silver from two different wound dressings in rats after a burn trauma. Wound Repair and Regeneration. 2014;22:654-659.

[77] Sengupta J, Ghosh S, Datta P, Gomes A, Gomes A. Physiologically important metal nanoparticles and their toxicity. Journal of Nanoscience and Nanotechnology. 2014;14:990-1006.

[78] Wei L, Lu J, Xu H, Patel A, Chen ZS, Chen G. Silver nanoparticles: synthesis, properties, and therapeutic applications. Drug DiscoveryToday 2015;20:595-601. doi: 10.1016/j. drudis.2014.11.014.